African Americans in the Civil War

African Americans in the Civil War

A Pictorial History of Courage and Pride

J. David Dameron

Southeast Research Publishing, LLC
An African American Experience Project

© 2017 by J. David Dameron

All rights reserved.

No part of this publication may be reproduced, stored in a retrieval system, or transmitted, in any form or by any means, electronic, mechanical, photocopying, recording, or otherwise, without the prior written permission of the publisher.

Printed in the United States of America.

Cataloging-in-Publication Data is available from the Library of Congress.

>ISBN 13: 9780692861431

>10 09 08 07 06 5 4 3 2

Picture Credits
Front: Cover image- Courtesy of the Library of Congress
Title: [*Unidentified African American soldier in Union uniform with wife and two daughters*] Call Number/Physical Location: AMB/TIN no. 5001 [P&P]; LC-DIG-ppmsca-26454. Source Collection: quarter-plate ambrotype; 13.9 x 16.4 cm (frame) Ambrotype/Tintype filing series (Library of Congress) Liljenquist Family collection (Library of Congress); Library of Congress Control Number: 2010647216; Repository: Library of Congress Prints and Photographs Division Washington, D.C. 20540 USA.

Summary: Photograph showing soldier in uniform, wife in dress and hat, and two daughters wearing matching coats and hats. In May 1863, U.S. Secretary of War Edwin Stanton issued General Order No. 143 creating the Bureau of U. S. Colored Troops. This image was found in Cecil County, Maryland, making it likely that this soldier belonged to one of the seven U.S.C.T. regiments raised in Maryland.

Published by
Southeast Research Publishing, LLC
An African American Experience Project

This book is respectfully dedicated to American Veterans, of all races, and all wars.

Contents

Acknowledgments		i
Introduction		ii
Chapter 1	*Slavery*	1
Chapter 2	*The War*	15
Chapter 3	*Mr. Douglass*	29
Chapter 4	*Emancipation*	43
Chapter 5	*Medal of Honor*	49
Chapter 6	*Unknown Spirits*	61
Chapter 7	*Courage and Pride*	75
Chapter 8	*The Nurses*	85
Chapter 9	*Mrs. Tubman*	99
Chapter 10	*Civil War Timeline*	111
Chapter 11	*Selected Speeches*	115
Chapter 12	*The Regiments*	169
Bibliography		175
List of Illustrations		179
Index		183

Acknowledgments

I owe a great deal of gratitude to many people, institutions of higher learning, libraries, archives and museums for their assistance in making this project a reality.

The National Portrait Gallery houses many of our national treasures in the form of paintings, engravings and photographs. I especially appreciate the assistance of Ms. Erin Beasley, who spent considerable time and effort in obtaining critical information and several rare photographs for this book- thank you!

I am also sincerely grateful for the kind assistance of the archivists at the National Museum of African American History and Culture. Their holdings are vast, and they contain a wealth of information. These national treasures remain readily accessible to everyone due to the hard work and diligence of these guardians of America's past.

I am very grateful to the Columbus Museum in Columbus, Georgia. Their archives hold many rare photographs and artifacts of Southern history. I especially thank Ms. Aimee Brooks for her assistance in obtaining the photographs of Mr. and Mrs. Horace King. I thank Christopher S. Morton, Assistant Curator of the New York State Military Museum and Veterans Research Center of Saratoga Springs, New York for his expert assistance. The NYS Military Museum holds a vast collection of rare Civil War era artifacts and images. I also thank Mr. Tom McCarthy for sharing his knowledge and photographic resources associated with the 26th United States Colored Troops regiment.

The National Underground Railroad Museums and Freedom Centers located in several locations throughout the United States were very helpful as well. There are Underground Railroad museums in: Tennessee; Kentucky; New York; Ohio; Pennsylvania; Maryland; and New Jersey. These wonderful organizations preserve and interpret the history of the conduit to freedom that so many African Americans traveled through and supported. I am very grateful for their assistance in getting the factual material about the Underground Railroad presented accurately.

The Donovan Research Library in Fort Benning, Georgia contains many rare publications. Their holdings are extremely rare as the library supports the U.S. Army's Maneuver Center of Excellence. This institution is responsible for training Infantry and Armor soldiers. Their archives are filled with U.S. Army operational journals, records, rare maps and imagery as well. The professional assistance rendered there by Genoa and Ericka is very much appreciated ladies- thank you!

The following historical institutions also provided assistance, and I thank them as well: the National Civil War Naval Museum; the National Infantry Museum at Fort Benning, Georgia; the Alabama Department of Archives & History in Montgomery, Alabama; the Georgia Department of Archives & History in Morrow, Georgia; the National Archives and the Library of Congress, both in Washington D.C.; the National Park Service in so many locations who selflessly maintain our National Battlefield Parks and Museums.

Finally, I must thank my lovely assistant and Chief Editor, Pamela Dameron. Her patience and loving support were critical throughout all phases of the manuscript preparation- thank you!

Introduction

The concept for *African Americans in the Civil War: A Pictorial History of Courage and Pride* was conceived while working on another project, a biography of Horace King. Mr. King's story is like the other African Americans portrayed in this book, he was born a slave during the nineteenth century, he lived through the Civil War, and when he died, he was free and beholden to no man.

During his life, Mr. King scaled the social ladder, and despite the racial inequalities of that era, he became a successful bridge builder and legislator. With great pleasure, a short biography of Mr. King is provided in this book and it is hoped that readers enjoy viewing his photograph as well. Along with the story of Horace King, this book also highlights ten other selected African Americans with images of each person as described.

Thus, the following people are highlighted herein as their stories present a varied perspective of the African American experience during the Civil War: Sojourner Truth; Horace King; Alexander Augusta; Anderson Abbot; Elizabeth Keckley; Frederick Douglass; Robert Smalls; Martin R. Delany; William H. Carney; Susan Baker King Taylor; Prince Rivers; Harriet Tubman; Hiram Revels; and Blanche Kelso Bruce. While there are many other interesting African Americans who contributed greatly to the struggle for equality and our American History, it would take volumes of books to highlight them all.

Additionally, room had to made to provide the presentation of numerous images of far too many "Unidentified" people. Nonetheless, their presence herein speaks loudly as their portraits reflect a people of courage and pride. As written by one of the people highlighted in this book,

We do not, as the black race, properly appreciate the old veterans, white or black, as we ought to. I know what they went through, especially those black men, for the Confederates had no mercy on them; neither did they show any toward the white Union soldiers. I have seen the terrors of that war.

- Susan Baker King Taylor

Sadly, there is a vast abundance of African American Civil War military photographs that remain labeled, "Unknown." Yet, the selfless, unassuming service of American war veterans has always been a hallmark of national pride and that is precisely what those men fought so hard to achieve. They weren't fighting for individual recognition, they fought for freedom and equality.

Hauntingly, their nameless imagery reflects a strong testament of selfless service, and they echo through time, with courage and pride. As time passes, families fade away, estates are sold and with no recollection of individuals we are left with only images listed in collections as "Unidentified" or "Unknown." Tragically, their names are forever lost to history.

Nonetheless, photographs are extremely important elements of history as they provide a frozen moment in time for future generations to view. After all, we are graphic creatures who rely upon our sense of vision for communications and we depend on these great insights for deeper comprehension. Anyone who possesses an important photograph and loses it, knows full well its value, once it is gone.

Viewing imagery associated with people and historical events that happened long ago, provides us with a far deeper and wonderful ability to better understand the past. Therein lies the purpose of this book, to highlight the story of African Americans during the worst internal violence the United States ever experienced, the American Civil War. By its very nature, a civil war is the most impassioned brutality and ugliest form of warfare known to man.

This book also delves into the root causes leading up to that horrific war, with some imagery that has heretofore, never been published. The book also provides not only imagery and stories of important African Americans, but also, those poor souls who lived and died in obscurity. I believe this imagery is an important national treasure and it has been ignored by scholars or simply not accessible to the public for far too long.

The photographs in this book were specially selected from several important public collections and in many cases, they were until very recently held in various private collections. This of course, is another reason why many of these images have never been accessible to the public.

I must explain to the reader that one of the largest and single most important, historically valuable, photographic contributions made to the American people occurred just within the last few years. The Library of Congress (LOC) in Washington D.C., houses a huge collection of documents, books, recordings, art work, and historical artifacts. The LOC depends a great deal upon the generosity of people to contribute items to what the library calls, "The American Memory."

The story of one such benevolent gift was a selfless act and quite touching. There is a link below to read more about this fantastic contribution to their fellow citizens. In brief, Mr. Tom Liljenquist and his family graciously donated seven-hundred photographs of Union and Confederate Civil War soldiers to the library in 2010. They continue to donate selected photographs to the library and they are all unique treasures. Mr. Liljenquist and his three sons collected these portraits over a period of several decades. All the images are interesting, high quality and as such, they invite viewers to look closely into these faces of the past. They do nothing less than bring history alive.

Several photographs from that collection are contained herein including the cover photo. Have you ever seen a photo of an African American family that was taken during the Civil War? They are extremely rare yet there are several of them presented herein. The father clearly exudes pride and he is resplendent in his Union uniform. Surrounding the soldier are his loving wife and daughters. The image is a gem and it provides an ultra-rare glimpse into lives of African Americans during the Civil War.

The viewer is quick to ponder, did the girls' Daddy live through the war? Did the soldier's wife and daughters survive the war? What were their names? Where were they from? In some cases, we know a few priceless details, but sadly, in most cases, their individual stories are lost to the ages.

Nonetheless, as we read from other documents and view similar photographs of that era, a greater comprehension is gained. Each person has a story and even when we don't have the complete details, one can view pictures and learn a great deal about the subjects therein.

Another wonderful collection of Civil War images housed in the Library of Congress is the William A. Gladstone Collection of African American Photographs. This collection was obtained by the library in 1995, and it is comprised of three-hundred-plus images of African Americans in various social settings and military periods, which includes the Civil War. In fact, the largest portion of the Gladstone Collection is from the Civil War era, and they're all African Americans.

The National Portrait Gallery and the Smithsonian museums in Washington D.C. have interesting photography collections also and a great deal of their photos are from the Civil War era. Several African American photographs herein were graciously provided from their collections. There are also many photographs housed in municipal, regional and state archives and museums. A good example is the Columbus Museum in Columbus, Georgia. Their collection has the photographs of Mr. and Mrs. Horace King, which you will find herein. Another example is the University of Toronto Museum in Canada. Their collection has the original and very rare photos of U.S. Army African American physicians, Alexander Augusta and Anderson Abbot. Their stories and images are presented within these pages as well.

Regarding photography, there are several important points to consider. The first photograph that included people was taken in France by Mr. Louis Daguerre in 1838. The photographic process he pioneered is known as the Daguerreotype, which was very popular in the United States during the Civil War. Once the science behind the permanent capture of images via chemistry was understood, various innovations and techniques were employed. By the beginning of the war, there were photographers in most of the major cities. More importantly, common people could afford to have their portraits made, and they did.

Several other popular variations of nineteenth century photography include Ambrotypes and Tintypes. Ambrotypes were made by underexposing a glass negative in front of a dark background. This causes a positive image to be created when exposed to light. Using a thin **layer of iodized collodion dipped in a silver nitrate solution,** Ambrotypes or the wet plate collodion process was quite popular amongst nineteenth century photographers.

For an additional fee, photographers would then add colored pigments to the plate for color effects. In this book, you will see where the artists added gold paint to buttons, buckles and jewelry. Some added tinted colors to other parts of the photo such as clothing to enhance the image as well.

Tintypes are like Ambrotypes but they employ a thin sheet of metal coated with dark paint used to produce a positive image. Tintypes were often hand-tinted as well. Ambrotypes and Tintypes were quicker, cheaper to produce, and more popular during the Civil War than Daguerreotypes. Daguerreotypes, Ambrotypes and Tintypes were all sold with various cases and elaborately designed frames which not only looked nice, they also protected the very fragile glass image.

These types of images cost between .25 cents to $2.50 depending on the photographer, location, size and the local market demand. In modern comparative costs, the photos would range today between $8.00 to 80.00 dollars for a 1/6 plate photo, roughly 2.5 x 3 inches. A common laborer earned around $2.50 a day, while soldiers earned between $7.00 to $20.00 per month, depending upon rank and race.

Several other innovative techniques were also employed using paper based images. These photos were known as cabinet cards or the "Carte de Visite." These thin card-type, paper portraits were roughly 4x 6 inches and quite often they appear as the "old-timey," sepia-toned images. As the paper products became popular in the 1860s, the photographers began to add elaborate framed lines and their logo is sometimes printed at the bottom and back of the card.

All these "albumen prints" as they were also called, were taken in portrait studios but there are some that were employed in landscape photography as well. These photographs were very popular by the end of the war and they were often placed in albums and cabinets for viewing. There are many examples of all these different types of photographs within this book. See if you can recognize the various types as you explore the pages.

The old cliché, "a picture is worth a thousand words" is certainly true. I encourage the reader to explore these topics in greater detail via the Library of Congress and National Archives website links below:

For an overview of various nineteenth century photography collections visit the National Archives overview at: https://www.archives.gov/research/alic/reference/photography.html#1800

Also, be sure and read more about the Liljenquist Collection in the Library of Congress article, "From the Donor's Perspective-The Last Full Measure". Available via the internet at: https://www.loc.gov/rr/print/coll/633_lilj_measure.html

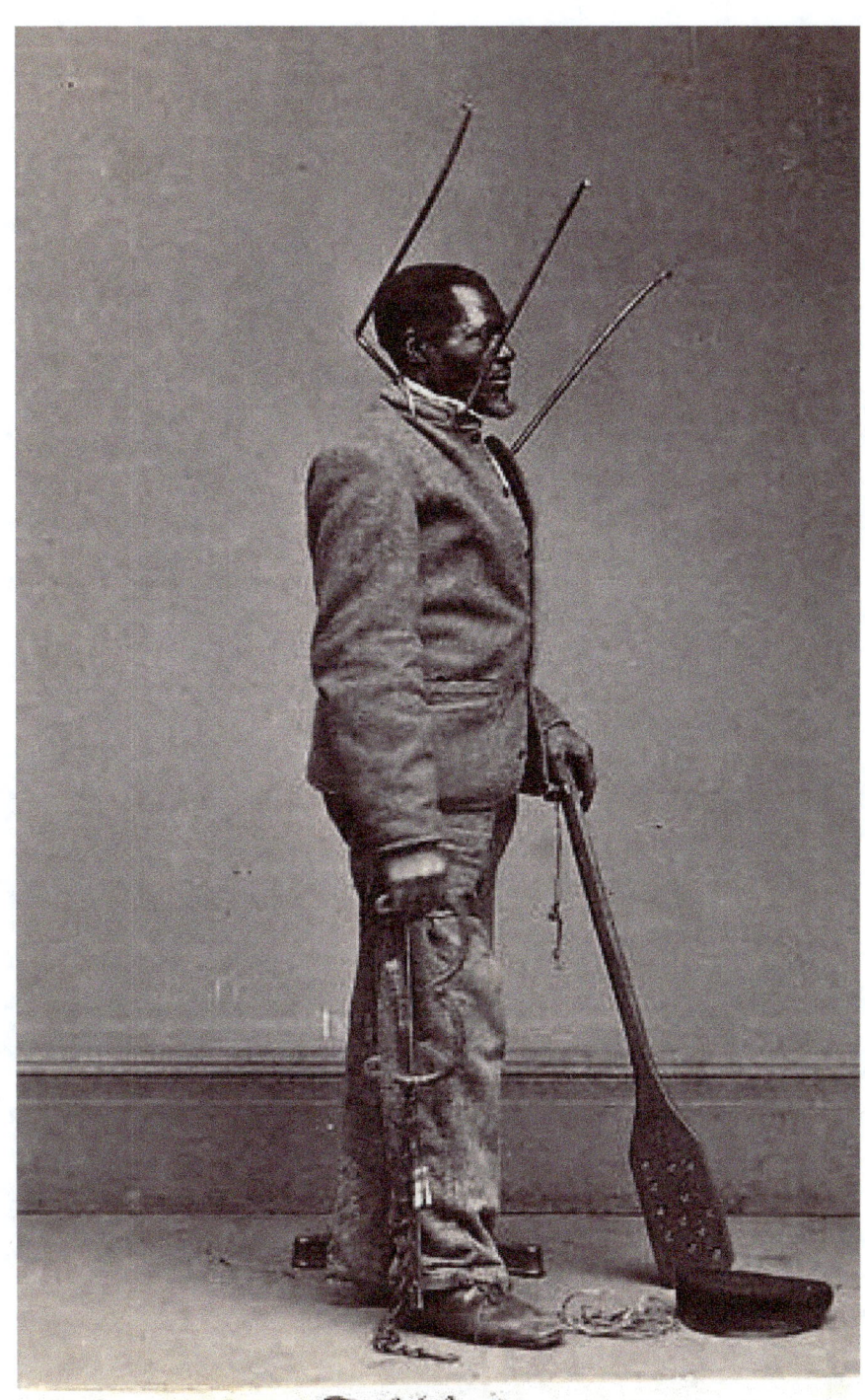

Wilson.

Branded Slave from New Orleans.

A branded slave and instruments of torture (GC)

"Our Countrymen in Chains."
An 1837 antislavery broadside publication of John Greenleaf Whittier's poem.
The design was originally adopted as the seal of the Society for the Abolition of Slavery
in England in the 1780s. It appeared on several medallions for the society made
by Josiah Wedgwood in 1787. (LOC)

Chapter 1

Slavery

"When I left the house of bondage, I left everything behind." – Sojourner Truth

Since the dawn of mankind, tragically, slavery has been a part of the human experience. Even today, these horrors are pervasive throughout the world. Known today as human trafficking, it is a multi-billion-dollar industry, and yes, it is evil. African Americans have suffered greatly from the tragedies of slavery and those dark chapters are forever etched in history.

Throughout the eighteenth and nineteenth centuries in the United States, the institution of slavery dominated the lives of African Americans. The importation of slaves from Africa had been a key part of the establishment of the nation. Africans were imported for inexpensive labor and they were put to work on Southern plantations as well as common laborers and servants in the North. As an institution, slavery was prevalent throughout the United States until 1789 when abolitionists in the Northern states began petitioning, by law, for its termination. Slaves migrated to these free Northern states prompting Southern slave owners to protest in Congress.

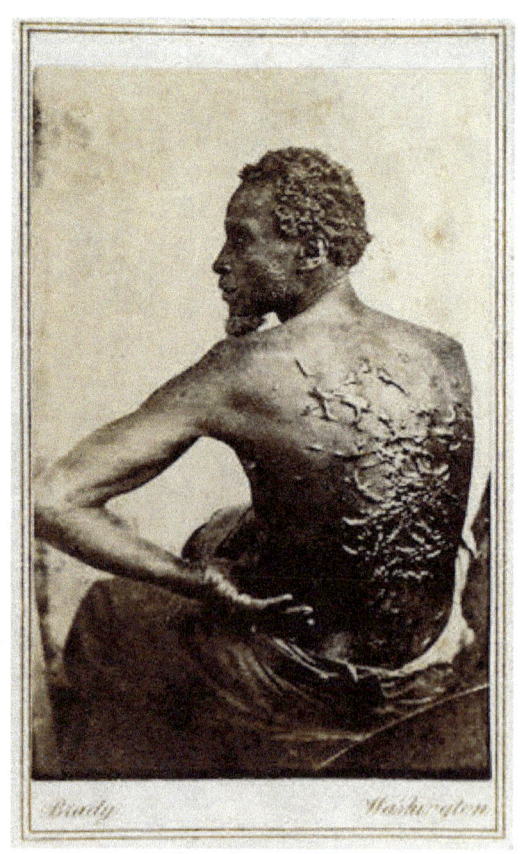

"Gordon" (NPG)

In 1793, the US Congress passed the Fugitive Slave Laws which gave protective rights to slave owners: the right to track down and reclaim escaped slaves; and the right to impose penalties on anyone who aided in their flight or provided fugitive slaves with sanctuary.

Many citizens were appalled at these laws. Abolitionist movements, churches, and politicians called for an end to the cruel and "peculiar institution" known as slavery. By 1821, the number of slave and free states each numbered twelve. By 1858, slave states had become a minority with slave states numbering fifteen and free states, seventeen.

This division was based primarily on economics as the Southern states relied upon the cheap labor of slaves to harvest crops (primarily cotton) and to perform the multitude of difficult manual tasks associated with large scale farming on plantations. In fact, slavery was a global practice but as the world modernized, the cruel nature of enslaving human beings spurred social change. By the early 1800s, slavery as an institution was made illegal in many nations. In the United States, both religious and political movements called for abolition, or an end to slavery.

Slave Families on a South Carolina Plantation. (LOC)

The Southern states insisted upon their rights as individual states and under the laws of states' rights they vowed to continue owning slaves. In 1850, the U.S. Congress enacted a series of laws known as the Compromise of 1850. These laws enhanced Fugitive Slave Laws by requiring government law enforcement officials throughout the nation to arrest and return runaway (fugitive) slaves. This law provoked a firestorm as "Northerners" refused to enforce these new laws.

Slaves planting sweet potatoes on a plantation in South Carolina (LOC)

    During the 1850s to 1860s, several legislative and supreme court decisions played key roles in the escalation of differences between the North and South. The Kansas-Nebraska Act, allowed settlers in those new territories to allow slavery within their borders. Violence erupted in the territorial regions and all too soon, Kansas was known as "Bleeding Kansas." The Supreme Court ruled in the Dred Scott decision that no one of African descent was qualified for U.S. citizenship. The nation was ripping apart as anger in abolitionists and slave owners swelled throughout the land.

Sojourner Truth- Isabella "Belle" Baumfree (NPG)

During this time, Sojourner Truth was a fervent abolitionist. Born, Isabella "Belle" Baumfree (1797-1883), she was a slave who became one of the first African American abolitionist. She also protested openly for women's rights (this was against societal norms). Born in New York at a time when slavery was practiced in the North, Ms. Baumfree grew into womanhood and developed a faith in God. She also rebelled against slavery. In 1826, she bravely fled from bondage with her daughter and returned to fight a legal battle against her former white master and won her son's freedom as well. Slavery was abolished by law in New York in 1827.

Mrs. Juliann Jane Tillman, an early female preacher and abolitionist of the African Methodist Episcopal (A.M.E.) Church (LOC)

In 1843, Ms. Baumfree, a devout member of the A.M.E. Church declared that God had called upon her and in her own words, she declared:

My name was Isabella; but when I left the house of bondage, I left everything behind. I wa'n't goin' to keep nothin' of Egypt on me, an' so I went to the Lord an' asked Him to give me a new name. And the Lord gave me Sojourner, because I was to travel up an' down the land, showing the people their sins, an' bein' a sign unto them. Afterwards I told the Lord I wanted another name, 'cause everybody else had two names; and the Lord gave me Truth, because I was to declare the truth to the people. - Published by Harriet Beecher Stowe, in "Sojourner Truth, the Libyan Sibyl," Atlantic Monthly, April 1863.

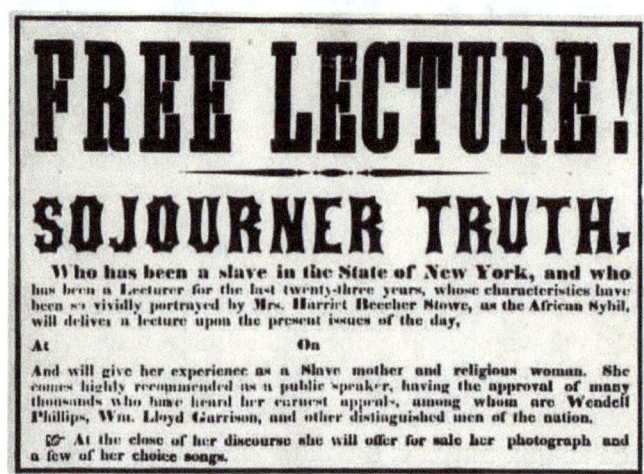

Sojourner Truth broadside advertising an appearance and a free lecture. (GC)

Sojourner spoke widely and she attended abolitionist and women's conventions where she rallied in support of equal rights. One of her most famous speeches was entitled, *"Ain't I a Woman?,"* which demanded equal rights for women as well as African Americans.

Later, during the Civil War, she recruited African American soldiers. Her grandson, James Caldwell, joined the U.S. Army and he served in the 54th Massachusetts U.S. Colored Troops. She also continued her quest to speak the truth and achieve equality for all. She was good friends with the abolitionists of her day and she was somewhat of a celebrity as well. Truth also worked in the hospital and at the Freedmen's relief organization in Washington D.C. In 1864, she was invited to the White House for an audience with the President.

On October 24, Truth and President Abraham Lincoln enjoyed a courteous and mutually respectful discussion about equality and Lincoln showed her a Bible given to him by the "colored people of Baltimore." After the war, Sojourner Truth continued her testimony and argued for equality for many years. Sojourner Truth passed away on November 26, 1883. Today, her legacy as a champion of abolition and equal rights echoes proudly through the ages.

President Lincoln showing Sojourner Truth the Bible presented by the "Colored People of Baltimore," Executive Mansion, Washington, D.C., Oct. 29, 1864.

In fact, on April 20, 2016 the U.S. Department of Treasury announced that the image of Sojourner Truth will be featured by 2020 on the newly minted $10.00 bills.

In the early nineteenth century, there can be found numerous references to Slavery as a "peculiar institution." The industrial Northern states participated in slavery just like their neighbors in the South, but they were not as reliant upon the institution as were the planters in the South. For many years, prior to the mass emancipation of African American slaves, people in polite society treated the topic gingerly as it was the political topic of that time.

Auction Slave House, Alexandria, Virginia (LOC)

Inside view of cells in a Slave House (LOC)

Unknown African American woman behind Slave House (LOC)

Slaves on South Carolina Plantation (LOC)

  During the early nineteenth century, the United States also underwent several waves of deep religious reflection. Known as the "Great Awakening," religion helped to highlight the wrongs of slavery in a moral society. Consequently, most states including those in the South, afforded slave owners with a voluntary, legal means of slave manumission. This legal means of emancipating slaves was very complex and owners released their slaves for many sentimental, benevolent, religious, and economic reasons. An example of manumission can be found in the story of Horace King.

  Horace King (1807-1885) was born into slavery on September 8, 1807, in Cheraw, South Carolina. His father was Edmund King, a Negro slave, and his mother was Susan King, a Catawba Indian, both of whom were the property of a local physician named King and the source of the family's surname. Upon his father's death, Horace was sold as property first to Mr. Jennings Dunlop, a slave trader, and then to his final owner, Mr. John Godwin, a builder.

As a young man, Horace King was educated to read and write at a time when educating blacks was viewed as unnecessary and in some places, it was even a crime. However, his new owner was a genteel man and he and Horace began a lifelong friendship that spanned many decades. Mr. Godwin decided that Horace would make a fine carpenter and so he educated him and put him to work.

John Godwin was a master builder and Mason. As Horace matured, he served his master as a construction laborer while absorbing his new skills and he excelled at everything he did. Horace adopted a saying that he often repeated to himself and others, "ignorance breeds poverty."

As an adult, Horace King gained his master's confidence as a trustworthy and energetic supervisor of his fellow slaves and together, in 1827, Godwin and King built a covered bridge that spanned the Pee Dee River in Cheraw, South Carolina.

Over time, King mastered bridge building and Godwin entrusted him to supervisory duties and relied on him more and more. King's abilities as a bridge builder and more importantly his ability to lead others provided him a dignified position and respected authority.

In October of 1832, Godwin moved his family and slaves west into Georgia as the former Creek Indian territories were opened to settlement. Godwin chose the bustling town of Columbus, Georgia. Shortly thereafter, they moved across the Chattahoochee River to Girard, Alabama, which is presently known as Phenix City, Alabama.

Horace chose a beautiful young local girl of mixed blood, Native and African American descent to be his bride. On April 28, 1839, Horace and Frances King were married and they quickly started a family. Ultimately, the King family was comprised of four sons and one daughter.

King worked as a slave until 1846, when his master, John Godwin petitioned the State of Alabama to emancipate him, and thus Horace King became a free man. King and Godwin continued working for many years and together as partners, they built numerous bridges, homes, churches, warehouses and by 1858, Mr. King achieved the status of Master Builder and Freemason. Mutual respect and a successful business venture forged powerful bonds of friendship between these men.

Together, Godwin and King built numerous bridges and several large commercial building projects. In 1859, John Godwin died and Horace King took care of all the arrangements including an expensive and beautiful marble obelisk, which states, "This stone was placed here by Horace King. In lasting remembrance of the love and gratitude he felt for his lost friend and former master."

Many years later, Robert Ripley wrote about King's admiration of his former master in his "Ripley's Believe It or Not," news article. Ripley wrote, "While their friendship may have seemed odd (due to race relations back then) to some people, clearly their life-long bonds are perpetually declared on that monument."

Mr. King employed his entire family in his construction business and his sons were excellent craftsmen, like their father. By 1860, Horace King was one of the wealthiest black men in the South.

In 1861, the Civil War erupted and although Mr. King was a free man, his sons faced conscription as laborers in the Confederate military. King secured an arrangement wherein his sons could work for their father and assist in the construction of war ships and other projects. The King family worked for the Confederate Navy at nearby, Port Columbus on the Chattahoochee River.

Mr. and Mrs. Horace King (Columbus Museum)

However, the stress of the war took its toll on King's wife and she died in 1864. After the war, Mr. King was persuaded to enter politics and he worked in local civic positions of authority and he was twice elected to the Alabama State Legislature. Meanwhile, he and his sons continued their construction business, which flourished during the Reconstruction Era.

In 1872, Horace King remarried and his family relocated to LaGrange, Georgia. In LaGrange, King's adult children assumed the primary responsibilities of bridge building and they formed a formal partnership under the business name of the "King Brothers Bridge Company." Under their father's supervision, the King family prospered.

Red Oak Creek Covered Bridge.
Constructed by Horace King in 1840 and still standing today. (GC)

On May 27, 1885, the master bridge builder died and he is buried in LaGrange, Georgia. His obituary in the local paper states that King had "risen to prominence by force of genius and character... we trust that the grace of God bridged for him 'the narrow stream of death' and that he now rests from all earthly cares and labors in the peaceful land beyond the flood."

Mr. King's legacy resides on a solid foundation of public service and engineering achievement. During his lifetime, Horace King participated in the construction of hundreds of bridges and countless buildings. It is amazing that despite the passage of time and weather, several of King's masterful constructions are still in use today. These architectural accomplishments bear witness to a genuinely skillful architect and engineer. Mr. King has been honored with many public accolades, and today, he is affectionately known as Horace "The Bridge Builder" King and the "Prince of Bridge Builders."

Horace King and Sons, Master Builders
(GC)

Stairs constructed by Horace King in the Alabama State House (LOC)

While Horace King was voluntary freed by his master, most Southern slaves were kept in servitude throughout their lives and they toiled under miserable conditions. Accordingly, abolitionists launched a fervent national movement during the 1850s to persuade people to abolish the institution of slavery.

A Ride for Liberty – The Fugitive Slaves, by Eastman Johnson (1862) (Brooklyn Museum)

An unidentified African American lady (LOC)

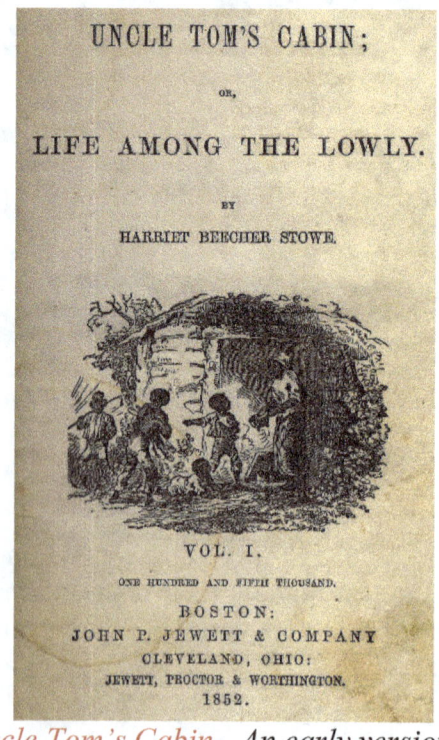

Uncle Tom's Cabin - An early version of Harriet Beecher Stowe's bestseller. (GC)

In 1852, *Uncle Tom's Cabin; or, Life Among the Lowly* by Harriet Beecher Stowe was published. This book was the best-selling novel of the nineteenth century, and second only to the Holy Bible as the most popular book during that era. In its first year of publication, it even sold a million copies in Great Britain. The people of Great Britain abolished slavery in 1833.

Most historians consider the book as one of the primary catalysts to the Civil War as it brought the brutal realities of Southern slavery to the forefront throughout the nation. Abraham Lincoln referred to Ms. Stowe as the "little lady who started" the Civil War.

The book's author was a teacher and abolitionist from Connecticut who wanted to highlight the evils of slavery and introduce readers to how blacks lived and suffered. The story is fictional and set in Kentucky but before the Civil War, even free Negroes in the Northern states were generally still treated as second class citizens. Still, the goals of the abolitionist movement were to abolish slavery and end institutional discrimination. To this end, Ms. Stowe's book was certainly successful.

Broadside advertisement (GC)

The book and plays, which are based on the book, stir emotions in people of the modern era as well due to the shocking mistreatment of a people based strictly on their race. Some point to the destructive stereotypes the author portrayed in the story's characters. This is especially true of the portrayals of elderly African American men as "Uncle Toms," the lazy, carefree "Happy Darky" young males, "picaninnies" as young black children and light-skinned mulatto women as loose sex objects. Some whites took offense to the role of Simon Legree as the cruel, greedy and overbearing white male antagonist of the story. Nonetheless, the book certainly played an important role in events of the nineteenth century and the Civil War.

Then, in 1859, violence resulted in a radical abolitionist raid on the U.S. arsenal at Harpers Ferry, Virginia (now West Virginia). Led by the fiery abolitionist, John Brown, he and twenty other men sought to ignite a slave revolt in the South but the plan failed. Brown and his accomplices were executed by federal authorities.

The census of 1860 shows a total population in the United States of thirty-one million people. Of those, four million slaves were held in bondage, primarily in the Southern states. In 1860, American voters also elected Abraham Lincoln as President. Violent reactions spread throughout the South, where the people called for secession from the Union. These states then formed their own government, the Confederate States of America (CSA).

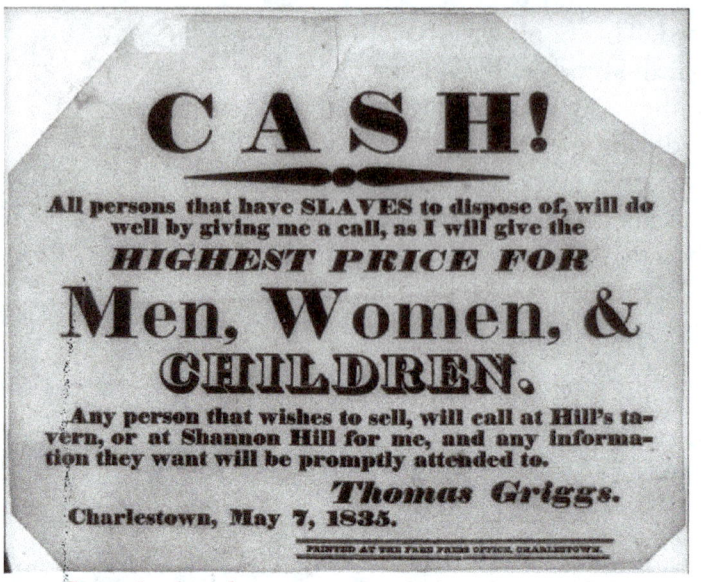

Broadside advertising Cash for slaves. (GC)

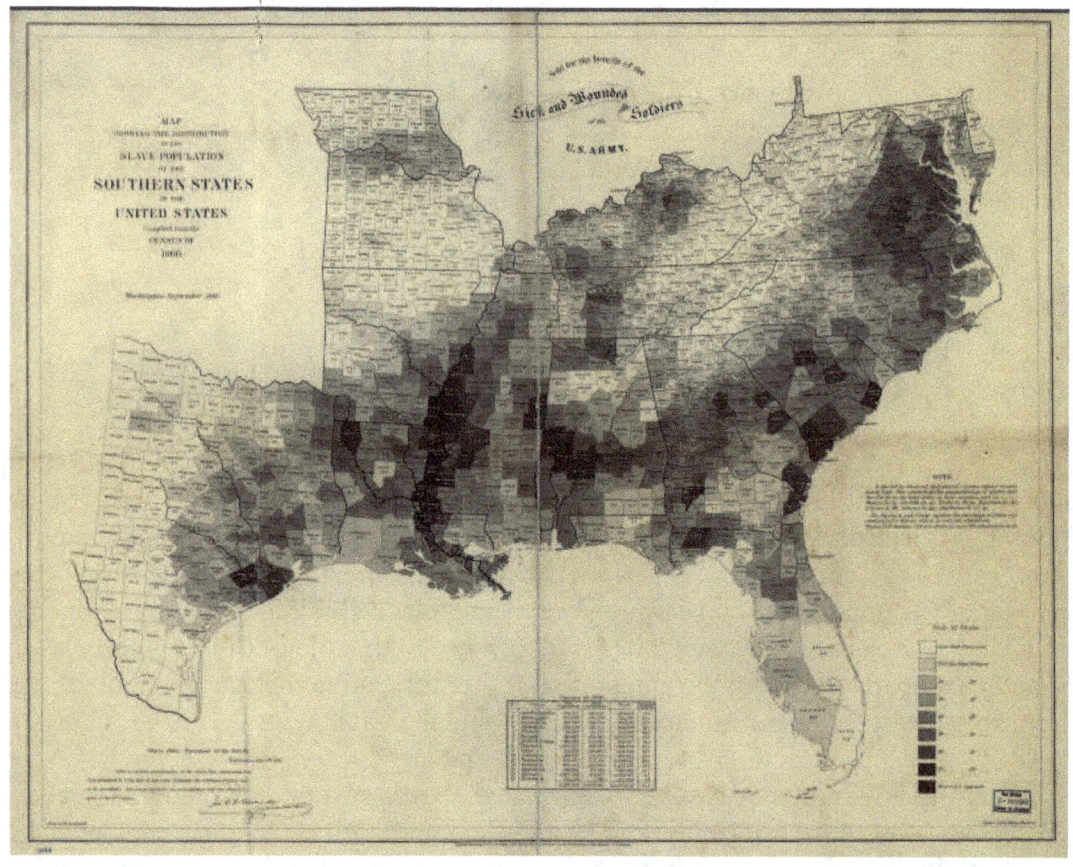

1860 Census Map - Slave populations in the Southern states. (U.S. Census Bureau)

Chapter 2

War

"Whenever I hear anyone arguing for slavery, I feel a strong impulse to see it tried on him personally." - Abraham Lincoln

Throughout the nation, the talk of war dominated conversations and it was clear to everyone that events were spiraling out of control. In the South, the CSA elected Jefferson Davis as their president and the new government prepared for war. In March of 1861, Lincoln took office and the U.S. Congress authorized the formation of a larger army in preparation for war.

On April 12, 1861, Confederates fired artillery barrages onto the federal fort in Charleston Harbor (Fort Sumter) in South Carolina, and the commander, U.S. Major General Richard Anderson was forced to surrender. On April 15, President Lincoln declared that the Southern states had embarked upon an active and illegal insurrection, and called upon the Northern state militias to be federalized into the Army. Between April and June of 1861, Virginia, Arkansas, North Carolina, and Tennessee also declared their secession from the United States. They too joined the Confederacy and the combined Southern states prepared for war.

As the nation was ripped into two separate governments, people were forced to choose their loyalties. As an example, Robert E. Lee was a federal soldier but he opted to join his native state of Virginia and he became a Confederate military advisor to President Jefferson Davis. In 1862, General Lee was appointed to the command of the Army of Northern Virginia, and ultimately, he was elevated to command the entire Confederate Army. Others had to make similar choices, and the war was soon considered an affair of "brother against brother."

For African Americans in the North, they were free men but at that time, they were restricted to service only as common laborers or in support positions. They were not allowed to serve in a combat role in the Army. In the South, African Americans were slaves or even if their masters had freed them, they were still forced to serve the Confederacy in some capacity as inexpensive laborers. Some joined alongside their masters and they went to war as personal servants or laborers but a few did voluntarily fight for the CSA. Many African Americans chose to flee northward and join the Union.

African American Teamsters- Union Army (LOC)

On the outskirts of Washington D.C., Camp Barker was created for housing and educating escaped slaves. Known in those days as "contraband," these men, women and children were housed and fed as they had no money and often only the clothes on their backs when they arrived from their arduous journey. President Lincoln assigned an African American surgeon, Captain Alexander Thomas Augusta (1825-1890) of the U.S. Army Volunteers to take charge of the camp.

Contrabands at Camp Barker, reading aloud in preparation for a visit from President Lincoln. (LOC)

Captain Augusta, had previously served as the unit Surgeon of the 7th United States Colored Troops. As a young freedman from Virginia, Alexander Augusta taught himself to read and write and he earned a living as a barber. He then moved to Maryland, Pennsylvania and on to California seeking entry into college but he was denied based on his race. So, in 1850, he then moved to Canada where he earned his way through medical school.

After six hard years of hard work and difficult studies, Dr. Augusta became a practicing physician, head of the Toronto City Hospital, and professor at the Toronto School of Medicine. While living in Canada, Augusta founded the Provincial Association for the Education and Elevation of the Colored People of Canada. This benevolent organization donated books and school supplies to black children. Shortly thereafter, he returned to the United States and when the Civil War began, he was commissioned as a surgeon in the U.S. Army.

During the war, he rose through the ranks and by the end of the war, he was a Brevet Lieutenant Colonel (one of the highest ranking black men in the military at that time). Augusta served as an administrator and medical advisor in military hospitals in South Carolina, Georgia, and Washington D.C. He was also the first black hospital administrator in U.S. history. After the war, in 1868, Augusta was the first African American appointed to the faculty of a medical college (Howard University) in U.S. history.

Captain Alexander Thomas Augusta,
U.S. Army Surgeon
Courtesy of the Online Image Bank
(University of Toronto)

Several times during his life, Augusta was persecuted and harassed because of hatred and jealousy, as he was a black man entrusted with positions of authority. Throughout his military service and later as a private citizen, Augusta fought hard against racial discrimination, writing in the newspapers and testifying in court and before Congress on behalf of equal rights. When Augusta died in 1890, he was buried in Arlington National Cemetery with full military honors. He was the highest ranking black soldier in the Civil War and the highest ranking black officer of that war buried in Arlington cemetery.

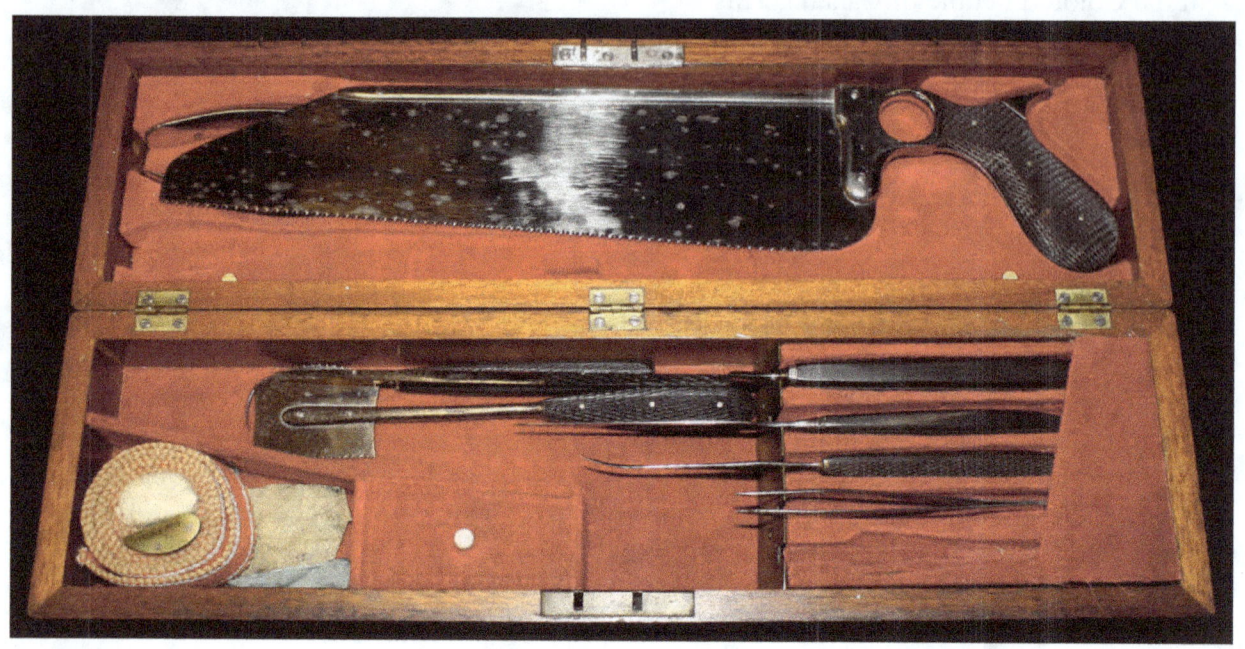

Surgeon's amputation kit. Although crude, the bone saw and other tools in the hands of a good surgeon enabled thousands of soldiers to survive the war.
Courtesy of the National Library of Medicine.

Anderson Ruffin Abbot (1837-1913) was the first Canadian-born physician of color. Anderson's family was originally from the United States (Mobile, Alabama) where they won their freedom and established a general store. Dissatisfied with racial persecution, they moved to Toronto, Canada and prospered. Anderson Abbot was born in Toronto and went to the school at the University of Toronto where he studied under the tutelage of Dr. Alexander Augusta.

In 1861, Abbot followed Dr. Augusta to the United States and volunteered for service in the United States Colored Troops (U.S.C.T). Abbot served under Dr. Augusta as a military physician in several of the Army hospitals administered by his former professor. While serving as a military surgeon in Washington D.C., Dr. Abbot became popular in society and he was also friends with the President and Mrs. Lincoln.

On April 14, 1865, President Lincoln was assassinated while watching a play at Ford's Theatre. Dr. Abbot and Mrs. Lincoln's African American dressmaker, Ms. Elizabeth Keckley attended to the First Lady as she mourned the loss of her husband. Mrs. Lincoln later showed her gratitude by presenting Dr. Abbot with several gifts, she gave him a plaid shawl and Scotch cap that had been worn by President Lincoln.

Captain Anderson Ruffin Abbot, U.S. Army Surgeon
Courtesy of the Online Image Bank
(University of Toronto)

Contrabands in a safe haven at Cumberland Landing, Virginia. (LOC)

Ms. Elizabeth Keckley (1818-1907) had served the first lady for four years. Originally from Dinwiddie Court-House, Virginia, Elizabeth was born a slave and served her masters' family as a domestic servant and nanny to his children. At the age of eighteen, she was traded to a family in North Carolina where she was sexually abused and soon she had a young son, George (Kirkland- his father's surname) to care for as well as the masters' family.

When she was still a young woman, "Ms. Keckley" as she was then known, was traded to another family who moved to St. Louis, Missouri where she enjoyed mingling with free blacks and earned money through sewing. Ms. Keckley yearned to be free and so she decided to borrow money from her friends and scrimped and saved money by selling dresses and doing seamstress work. In 1855, although it took her thirty-seven years, she convinced her masters to free her in exchange for her life savings of $1200 dollars.

Ms. Keckley continued working and paid back her friends and debtors the money she borrowed for her freedom. She also managed to send her son George to Wilberforce University in Ohio, the first college owned and operated by African Americans in the United States. Ms. Keckley's business as a dressmaker and seamstress did quite well. In 1860, she decided to move to Baltimore, Maryland.

As a seamstress, Ms. Keckley managed to earn $2.50 a day, but she wanted her own business. She then moved to Washington D.C. where she began doing dressmaking work for the elite ladies of Washington society. Her reputation spread amongst the ladies and within a few months, she was making dresses for Mrs. Robert E. Lee and Mrs. Varina Davis. Mrs. Davis later served as the First Lady and wife of President Jefferson Davis (CSA).

Engraving from the Frontispiece of Ms. Keckley's book, Behind the Scenes, published in 1868.

Matthew Brady took this photograph of Mrs. Lincoln. The gown was created by her Modiste and confidante, Ms. Elizabeth Keckley. (LOC)

Shortly thereafter, in 1861, Mrs. Mary Todd Lincoln having heard of Ms. Keckley's work, summoned her to the White House. After interviewing her, Ms. Keckley was selected by Mrs. Lincoln to be her personal modiste, a fashionable dressmaker of that era. Suddenly, Ms. Keckley was making a fortune as Mrs. Lincoln spent a great deal of money on her dresses, hats, and shoes. Ms. Keckley stayed busy making dresses and evening gowns for the numerous presidential events such as dinner parties and balls. In 1862, the famous photographer Matthew Brady took pictures of the first lady in several of Ms. Keckley's gowns.

In 1862, Ms. Keckley founded the Contraband Relief Association (CRA) to provide a lifting hand to many poor blacks who escaped enslavement in the South and traveled north seeking freedom. With no education or money, the former slaves needed considerable assistance. As the violence of the Civil War escalated and African Americans began serving in the U.S. military, Ms. Keckley expanded the CRA and renamed the organization the "Ladies' Freedmen and Soldier's Relief Association." Men such as Frederick Douglass, the champion of African Americans in his day, rallied to support Ms. Keckley's organization.

Thousands of former slaves or contraband as they were known reached safety in the North with little more the clothes on their backs. (LOC)

While Ms. Keckley sought to provide food, shelter and clothing for former slaves, she also warned them with a heavy dose of reality. The following passage is from her autobiography:

The North is not warm and impulsive. For one kind word spoken, two harsh ones were uttered; there was something repelling in the atmosphere, and the bright joyous dreams of freedom to the slave faded--were sadly altered, in the presence of that stern, practical mother, reality. Instead of flowery paths, days of perpetual sunshine, and bowers hanging with golden fruit, the road was rugged and full of thorns, the sunshine was eclipsed by shadows, and the mute appeals for help too often were answered by cold neglect.

Poor dusky children of slavery, men and women of my own race--the transition from slavery to freedom was too sudden for you! The bright dreams were too rudely dispelled; you were not prepared for the new life that opened before you, and the great masses of the North learned to look upon your helplessness with indifference--learned to speak of you as an idle, dependent race. Reason should have prompted kinder thoughts. Charity is ever kind. – Elizabeth Keckley, *Behind the Scenes*

The brutal violence of war coupled with the continuing racial bigotry Ms. Keckley encountered (even in the North), prompted her to become more vocal regarding race and equality. She also befriended other free blacks such as the African American military surgeons, Dr. Anderson Abbot and Dr. Alexander Augusta and she worked very closely with them to care for wounded soldiers. On several occasions, she and her organization prepared and hosted dinners honoring the men who served their nation, black and white. She also worked very closely with independent black churches and helped organize community concerts, fundraisers and speeches to raise money for relief activities.

Ms. Keckley also spent a great deal of time caring for the Lincolns' young children William (Willie) and Thomas (Tad). Her motherly instincts and domestic experience helped Mrs. Lincoln a great deal at a time when she was suffering from mental exhaustion. In 1862, both Willie and Tad suffered with Typhoid Fever. While Tad eventually recovered from the illness, little Willie died on February 20, 1862.

Mrs. Lincoln was inconsolable with grief but Ms. Keckley helped her through the pain of her great loss. Ms. Keckley's son George served in the Union Army and he had died several months previously at the Battle of Wilson' Creek, Missouri. Consequently, the ladies consoled one another and confided very closely. Ms. Keckley and Mrs. Lincoln forged bonds during the war years that would last the rest of their lives.

Willie and Tad Lincoln (at right), sons of President Abraham Lincoln, with their adult cousin, Lockwood Todd. (LOC)

When President Lincoln was assassinated, the distraught first lady gave Ms. Keckley many of her articles of clothing. Mrs. Lincoln even gave Ms. Keckley the blood-stained velvet cloak and bonnet she had worn to the play the evening of her husband's assassination. Ms. Keckley donated the clothing and memorabilia from the White House to Wilberforce College. The newspapers reported this as the "Old Clothes" scandal, which hurt Ms. Keckley very deeply. Mr. Lincoln, who was already suffering from depression, took the actions of Ms. Keckley and the negative publicity very seriously. Accordingly, their relationship became quite strained.

An artist's rendering of the Assassination (GC)

Then, in 1868, Ms. Keckley published her autobiography entitled *Behind the Scenes*. Again, the newspapers and society in general reacted negatively. Many people considered Ms. Keckley's book as taking great liberties in the recording of private matters and observances from her experiences in the White House. Ms. Keckley also recorded the rape at the hands of her former master, which was considered scandalous and unacceptable in the nineteenth century.

Accordingly, Ms. Keckley's dressmaking business suffered greatly and she was shunned by society. Thus, she moved away from Washington and she accepted a position in Ohio at Wilberforce University as the Department Chair of Sewing and Domestic Science Arts.

A pre-war lithograph of the Wilberforce University campus in Xenia, Ohio. (LOC)

In 1893, Ms. Keckley displayed an elaborate dress exhibit at the Chicago World's Fair. Ultimately, she moved back to Washington but she never regained the position in society she once enjoyed. Sadly, she had to move into the "National Home for Destitute Colored Women and

Children" where she died in May of 1907. The National Home was created by charity funds that she led in establishing so many years before.

Although the Lincoln Collection in the Smithsonian's American History Museum has several of Ms. Keckley's gowns worn by Mary Todd Lincoln, Elizabeth Beckley is buried in an unmarked grave in National Harmony Memorial Park in Largo, Maryland.

Two African American ladies. Fashionable, period dress, circa 1860. (LOC)

Slave advertisement- Great Negro Mart of Memphis (GC)

Underground Railroad Map highlighting the varied and secretive paths to freedom. (GC)

Chapter 3

Mr. Douglass

"You have seen how a man was made a slave; you shall see how a slave was made a man."

– Frederick Douglass

Frederick [Douglass] Augustus Washington Bailey (1818-1895) was born a slave on a plantation in Talbot County, Maryland. As a young field hand, Frederick suffered under a cruel master, Mr. Edward Covey who whipped him weekly to break his will. Mr. Covey was proud of his local reputation as a "negro-breaker." During one of weekly beatings, Frederick Bailey fought back and that stopped Mr. Covey's assaults.

Frederick tried several times to escape but he did not meet with success. While working as a leased laborer and ship caulker for his master in Baltimore, Frederick met and befriended Miss Anna Murray (1813-1882). Miss Murray secretly served as an asset in the Underground Railroad, which helped slaves in their quest for freedom. Frederick and Miss Murray fell in love.

Miss Murray was a free black and she arranged an escape plan for Frederick. Miss Murray arranged all the details for his escape. Armed with travel money, a train ticket and a disguise as a sailor, on September 3, 1838, Frederick achieved his freedom. Frederick traveled by train to Delaware, Philadelphia, New York and finally he achieved safety in Massachusetts. Two weeks later, Frederick married Miss Anna Murray. Initially they adopted the surname Johnson but later changed it to Douglass to ensure his freedom.

Frederick Douglass (NPG)   *Anna Murray Douglass* (LOC)

Mr. Douglass learned to read and write, and he preached in church, gaining invaluable skills as an orator. Mr. and Mrs. Douglass joined the African Methodist Episcopal Zion Church. Its membership in the nineteenth century also included Harriet Tubman and Sojourner Truth. The church broke from the main Methodist organization in 1821 because of overt discrimination against African Americans. This independent religious denomination remains a powerful African American church to this day.

His passion grew as an abolitionist, social reformer, writer, orator, and statesman. He also devoted his life to the quest of racial equality and he served for many years as a national leader of African American causes. He also helped women and Native Americans in their fight for equal rights as well.

Mr. Douglass wrote several autobiographies, *Narrative of the Life of Frederick Douglass, an American Slave* (1845), and *My Bondage and My Freedom* (1855), and *Life and Times of Frederick Douglass* (1881-revised in 1892). His books are filled with his own unique and forceful narratives that were intended to get peoples' attention, and they did.

In his day, witnesses stated that Mr. Douglass was so articulate and intelligent that it was hard to imagine he had been a slave. It is also said that he spoke with a powerful baritone voice and he seldom smiled in public. His serious nature was intended to convey determination, and it worked. His writing reflected a similar tone. As an example, in his first book, he defiantly declared, "You have seen how a man was made a slave; you shall see how a slave was made a man."

Mr. Douglass certainly served as an inspiration to his fellow African Americans. He used his talents wisely and achieved a great deal as the voice for a people who had known nothing but persecution and sub-human treatment for many years. Some historians describe him as the most influential and important African American of the nineteenth century.

Frederick Douglass. Engraving from *My Bondage and My Freedom.* (1855-GC)

Mr. Douglass often referred to the employment of peaceful agitation to achieve a desired societal change. As an example, he stated:

We have what is most valuable to the human race generally. It is a revelation of a power inherent in human society. It shows what can be done against wrong in the world, without the aid of armies on the earth or of angels in the sky. It shows that men have in their own hands the peaceful means of putting all their moral and political enemies under their feet and of making this world a healthy and happy dwelling-place, if they will but faithfully and courageously use these means.

In 1841, Mr. Douglass met William Lloyd Garrison a fervent, white abolitionist. Douglass considered Garrison's words second only to those of God and the Holy Bible. This huge influence on Douglass encouraged him to speak more openly in public against slavery and the goodness of the Underground Railroad. In 1845, he published his first autobiography about his experiences and beliefs, which met with huge success throughout the North.

In his books and speeches, Mr. Douglass always spoke freely and honestly regarding his thoughts. In one such thought he reflected on slavery with**, "***There is no negro problem. The problem* *is whether the American people have loyalty enough, honor enough, patriotism enough, to live up to their own constitution."* His words struck a chord with the American people.

Mr. Douglass and his friends began to fear for his life as public hatred also reared its ugly head due to his openly abolitionist views. Mr. and Mrs. Douglass moved to England and Ireland where he lived for two years and he was well-received by the people. Slavery had ended in Great Britain in 1838. Later recorded in his autobiography of 1855, Mr. Douglass recorded that while in Europe, *"I am seated beside white people—I reach the hotel—I enter the same door—I am shown into the same parlor—I dine at the same table—and no one is offended... I find myself regarded and treated at every turn with the kindness and deference paid to white people. When I go to church, I am met by no upturned nose and scornful lip to tell me, 'We don't allow niggers in here!'*

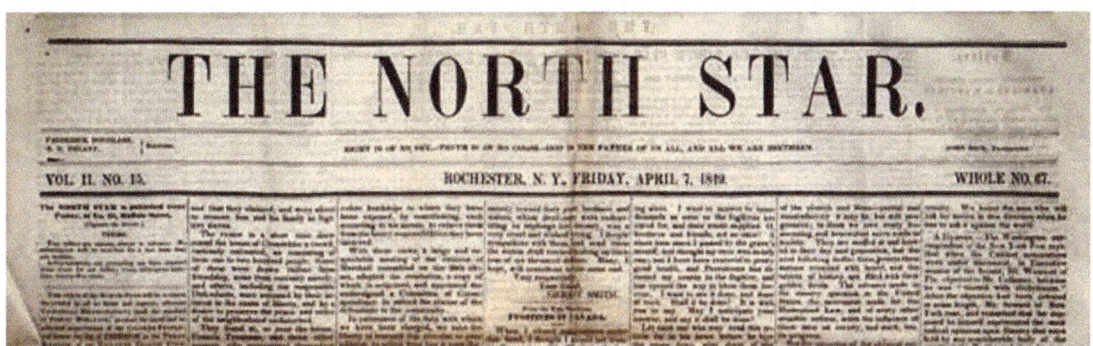

Periodical published by Frederick Douglass and his business partners, William R. Delany and William Lloyd Garrison before the Civil War. Founded in Rochester, New York, in 1847, the slogan of the North Star was, "Right is of no Sex--Truth is of no Color--God is the Father of us all, and we are all Brethren." Mr. Douglass fought not only for the abolition of slavery, as evidenced in his newspaper, he fervently supported women's suffrage and education as well. (GC)

Legally, Mr. Douglass was still a fugitive and while living in Great Britain, Mr. Douglass addressed his legal, fugitive status, from overseas. His friends in the United States arranged to formally pay his owners in Maryland, and thus they secured the freedom of Frederick Douglass. After returning to America, Mr. Douglass attended the Seneca Falls Convention in New York where he rallied with Elizabeth Cady Stanton in 1845 for the cause of equal rights for women. In 1852, Frederick Douglass delivered a powerful speech to the ladies of the Rochester, New York Anti-Slavery Sewing Society. Known as the "*What to the slave is the 4th of July*" speech, Mr. Douglass focused on the positive principles of American values (citizenship and freedom) and he then highlighted how none of these very important rights of man apply to African Americans.

In 1859, Mr. Douglass was encouraged to meet with the Radical abolitionist John Brown who told Douglass of his desire to take up arms against the government and hoping to start a war. Mr. Douglass refused any of Brown's desires to collaborate in violence and he fled to Canada fearing of guilt through association. Later of course, John Brown and his fellow conspirators were killed when they raided the federal military arsenal at Harpers Ferry, Virginia (now West Virginia).

In 1861, when the Civil War began, Mr. Douglass was widely known throughout the United States for his abolitionist views. After the 1st Battle of Bull Run, in which the Union suffered a major loss, Douglass wrote that he witnessed black soldiers fighting for the Confederacy and encouraged Abraham Lincoln and the people of the United States to allow African Americans to bear arms as well.

BLACK NATIONAL DEFENDERS!

The State of Connecticut is authorized to raise Colored Troops; and any number of her quota of 5,000 may be colored men. 29th Regiment Connecticut Volunteers, is now being formed at Camp Buckingham, composed entirely of Colored Men, located at the beautiful City of New Haven, the seat of Yale University.

STATE BOUNTY, $200.00 CASH!

On being sworn in.

By an old law of the State, 30 dollars a year are allowed to each soldier for clothing, 10 dollars of which is paid down at the time of entering the service, the other 20 dollars being paid in four month payments each, making 210 dollars Bounty—cash, on joining the Regiment—and 20 dollars more during the year.

An important fact connected with this recruiting is, that the contract for raising the troops has been given to a Colored Man; and Connecticut is the first State, since the war commenced, which has been thus liberal and considerate.

This fact alone should be an inducement for **COLORED MEN** to rally to her standard: all the Recruiting Agents in the West being Colored; and this principle should prevail everywhere. Colored Men should recruit Colored Men, as best adapted to it.

The most liberal compensation will be given to Good Agents, about 50 such being now wanted, and to whom will be paid Cash so soon as service is rendered.

APPLY WITHOUT DELAY TO
DR. M. R. DELANY,

State Contractor, Head-quarters of the West and South-Western States and Territories, 172 Clark Street, Top Story, Chicago, Ill.
JOHN JONES, Assistant.
LIEUT. W. F. STAINES, Evansville, Ind.

☞ For further information, address Dr. M. R. Delany, Box 764, Chicago, Ill.

NOTE.—I may add here that I am much indebted for obtaining this contract to Major Sanford, of Heavy Artillery Service, and other gentlemen in the Rhode Island Heavy Artillery, who recommended me to the authorities of Connecticut. M. R. DELANY, State Contractor.

Chicago, Ill., Dec. 1st, 1863.
P.S.—All recruiting in the States in which it is prohibited is hereby forbidden.
M. R. D.

Chicago Evening Journal Print, 60 Dearborn Street.

U.S. Army African American "Call to Arms" Recruiting Poster (NARA)

By March of 1863, dozens of United States Colored Troops (U.S.C.T.) had formed federal regiments and by May of that year they began deploying to battlefield areas for combat service. Mr. and Mrs. Douglass had five children, three sons and two daughters. Their sons, Lewis and Charles joined the 54th Regiment Massachusetts Volunteer Infantry, and Frederick Douglass Jr. (1842-1892) joined the 25th US Colored Infantry regiment.

Lewis Henry Douglass (1840–1908), his eldest son, quickly rose through the ranks and he served as the unit's Sergeant Major (the highest enlisted man in the unit). Charles Remond Douglass (1844 – 1920), the youngest son also joined the 54th Massachusetts but he became sick and could not deploy with his unit. Consequently, he transferred to the 5th Massachusetts Cavalry and ultimately to the Capital City Guards' Battalion due to continuing illness. Nonetheless, he became an officer and attained the rank of Major by the end of war. Even Frederick Douglass Sr. served his nation as a recruiter.

Charles Remond Douglass, 5th Massachusetts Cavalry (GC)   *Sergeant Major Lewis Henry Douglass, 54th Massachusetts* (GC)

In 1863, Frederick Douglass conferred with President Lincoln regarding African American soldiers and other racial matters. The president received the stern but respectful message of Mr. Douglass graciously, and undoubtedly, President Lincoln learned a great deal from the sage advice of Mr. Douglass.

After the war, Mr. Douglass was considered by most people as the unofficial leader and spokesman of black America. Frederick Douglass continued his fight for racial equality and the 13th (abolition of slavery), 14th (defines citizenship and equal protection) and 15th (right to vote) amendments to the U.S. Constitution were victories for all African Americans.

In 1872, Frederick Douglass was placed on the election ballot as the first African American nominated for vice president of the United States. While Mr. Douglass never campaigned for the position on the Equal Rights Party ticket, nonetheless, this was the first time in American History that an African American appeared on a presidential ballot.

During the Reconstruction Era, Douglass also championed the rights of women. Sadly, during this time, Anna Douglass died on August 4, 1882. They had been married for forty-four years.

In 1889, President Benjamin Harrison appointed Frederick Douglass to the post of minister-resident and consul-general to the Republic of Haiti. Mr. Douglass held that post for two years and as such, he was the first African American to serve his nation abroad as a statesman. On February 20, 1895, Douglass participated in the National Council of Women in Washington, D.C. where he was recognized for his selfless service. That evening, he died at the age of seventy-seven.

Many historians have recorded that Frederick Douglass was the most influential African American of the nineteenth century. While his burdens were many, they made his accomplishments even greater. In fact, his awards and honors are lengthy and his life works are the subject of many celebrated books and movies.

Frederick Douglass (NARA)

In addition to the sons of Frederick Douglass serving in uniform as thousands of other U.S. Colored Troops, a multitude of African American support troops were employed as cooks, servants, trench and grave diggers. In the U.S. Navy, that service allowed able bodied seamen of any race to serve aboard ships at sea and in the inland waterways but their duties were primarily manual labor. However, some African Americans did serve as ship pilots in the both the U.S. and Confederate naval forces.

One such pilot was Robert Smalls (1839-1915) of Charleston, South Carolina. Smalls was a slave originally from Beaufort but at age twelve, his master sent him to Charleston where he worked on the docks as a stevedore or dockworker. As he matured, he was trained as a ship's "Wheelman" as slaves in the Confederacy were not allowed to be called a ship "Pilot." As an adult, Smalls honed his abilities and piloted ships at sea and around the southern coastline. More importantly, he knew Charleston harbor like the back of his hand.

In 1861, the Civil War erupted at Fort Sumter in Charleston harbor, and Smalls was assigned to slave duties for the Confederacy as a "wheelman" of the CSS Planter, a large sidewheel steamship that had been converted into a rebel gunboat and transport.

On May 12, 1862, the CSS Planter docked at Charleston and the Confederate Captain and his white crew spent the evening ashore while the slaves maintained the ship. In the early morning darkness of May 13, 1862, Smalls and seven slave crewmen donned confederate uniforms and they quietly slipped the boat into the harbor. The men then stopped at a nearby dock where as previously arranged, they picked up their awaiting families and embarked on a stealthy mission.

With signal codes in hand, Smalls and his "pirate" crew sailed right past the Confederate fortifications just as ordinary as any routine operation. Smalls then continued directly towards the Union blockade. As he approached the Union Navy, Smalls attached a white bedsheet to the ship's mast indicating surrender. The Union Navy welcomed Smalls and his band of merry passengers to safety and freedom. The U.S. Navy secured the ship, cannons, ammunition and naval stores.

Robert Smalls and the *C.S.S. Planter* (Harper's Weekly)

Heralded throughout the press as a hero, Smalls was rewarded by the Union with an appointment as a U.S. Navy ship's pilot. Captain Smalls also received thousands of dollars for the ship he stole from the Confederacy. Then, he was dispatched to Washington D.C. where Captain Smalls influenced President Lincoln's decision to allow African Americans greater latitude in serving the U.S. military. Following his short-lived fame in the press, Captain Smalls returned to his duties.

In 1863, he decided to make the war more personal and he joined the Army where he served in numerous battles, including the bloody Battle of Fort Wagner, for the remainder of the war. Mr. Smalls survived the war and he served in the South Carolina legislature from 1868 to 1875, and the

U.S. House of Representatives from 1875 to 1887. Upon his monument, his own words are etched in stone, which declares, "My race needs no special defense, for the past history of them in this country proves them to be the equal of any people anywhere. All they need is an equal chance in the battle of life."

Following Smalls visit to Washington D.C, on July 17, 1862 the United States military authorities decided to officially allow Negroes to join the United States Army. Still, white officials would not allow an integrated land force. On August 25th, 1862 Secretary of War, Edwin Stanton authorized the formation of the first African American unit, the 1st South Carolina Infantry Volunteers, which was led by white officers.

The 29th Connecticut Colored Infantry. Beaufort, South Carolina. (LOC)

These troops were comprised of runaway slaves from Florida and the sea islands of the South Carolina coast where the U.S military had established a foothold deep in Southern territory, just south of Charleston. As these men rushed to safety in the US army camps, they eagerly volunteered to fight the confederates.

Men of the 107th Colored Infantry Regiment (LOC)

During the winter of 1862, pressures were building on the president. Between Douglass, the other abolitionists, the war, and the death of his son William, coupled with the wartime suffering of so many souls had a huge influence on Lincoln's decisions. He prayed and he weighed his options regarding African Americans. On December 1, 1862, Lincoln reported to Congress stating, "In giving freedom to the slave, we assure freedom to the free - honorable alike in what we give, and what we preserve. We shall nobly save, or meanly lose, the last best hope of earth. Other means may succeed; this could not fail. The way is plain, peaceful, generous, just - a way which, if followed, the world will forever applaud, and God must forever bless."

Meanwhile in the Confederacy, President Davis issued a proclamation, General Order #111 on December 24, 1862, which warned that white officers of Union African American troops captured on the battlefield would be put to death. He further declared that any African American soldiers captured on the battlefield would be remanded to the Southern states and dealt with per the laws of those states. The implication of this threat was that insurrectionist slaves were put to death in the South.

Indirectly, President Davis helped President Lincoln make up his mind and on January 1, 1863, he issued the Emancipation Proclamation as an Executive Order and United States "Colored Troops" (U.S.C.T) units began forming in large numbers throughout the Northern states.

President Abraham Lincoln was a hero of many and a very popular man in most African American families. (LOC)

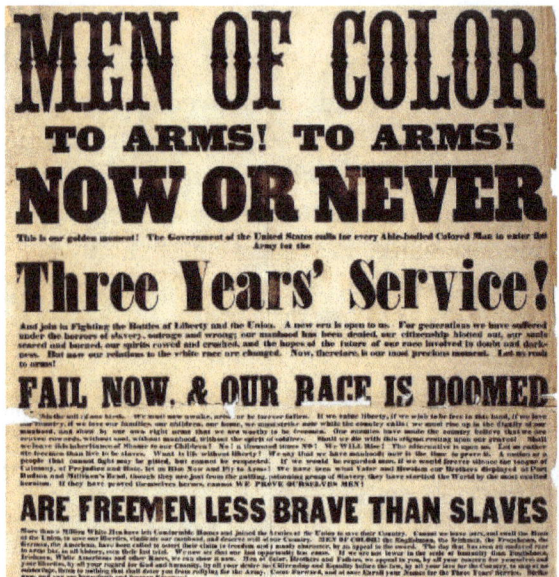

U.S. Army Recruiting Broadsides (GC)

More importantly, the proclamation decreed that millions of slaves, throughout America and even those held in Confederate States were henceforth, free people. In the Southern states, the proclamation was laughed off or ignored by slave owners as irrelevant political desperation. However, the word spread like wildfire across the land and slaves fled northward in droves. The Confederate authorities countered the claims of the US government and anger intensified throughout the nation.

Contrabands migrating northward and reaching Union lines. (LOC)

"First Reading of the Emancipation Proclamation."
Painting by Francis Bicknell Carpenter- 1864. (U.S. Census Bureau)

Chapter 4

Emancipation

"With broken hopes-sad devastation; A race resigned to DEGRADATION! - Martin R. Delany

In the camp of the 1st South Carolina Volunteers U.S.C.T., a grand celebration of the New Year and the Emancipation Proclamation was held on January 1, 1863. At a huge banquet and unit flag ceremony, emotions ran high. In the words of their Surgeon, Dr. Seth Rogers:

After the presentation speech, had been made, and just as Col. Higginson advanced to take the flag and respond, a negro woman standing near began to sing "America", and soon many voices of freedmen and women joined in the beautiful hymn, and sang it so touchingly that everyone was thrilled beyond measure. Nothing could have been more unexpected or more inspiring.

The President's proclamation and General Saxton's New Year's greeting had been read, and this spontaneous outburst of love and loyalty to a country that has heretofore so terribly wronged these blacks, was the birth of a new hope in the honesty of her intention. I most earnestly trust they not hope in vain… None of us had ever heard them sing America, and the most infinite depth and tenderness of "My country 'tis of thee Sweet land of Liberty," was inspiring to the last degree. (letter from M.D. Seth Rogers, January 1-2, 1863).

Watch meetings were held throughout the North in anticipation of President Lincoln's "hour of emancipation." Formally announced on January 1, 1863, African Americans were formally freed from the bonds of slavery. Finally, they were, "Free at last, Free at last!"
(LOC)

Another African American renowned for his brilliance was Martin R. Delany (1812-1885). Delany was born in Charles Town, Virginia (now Charleston, West Virginia). His father was a slave and his mother was a free black. She taught young Martin the importance of literacy, and he excelled in his studies. Seeking full freedom and a less hostile place to live, the Delany family moved to Chambersburg, Pennsylvania.

In 1831, at the age of nineteen, young Mr. Delany moved to Pittsburgh where he worked as a laborer and barber. Martin Delany desired to visit Africa as family folklore taught him that his ancestors were tribal royalty. He was also becoming very upset that in his view, whites were restricting blacks from opportunities. This would later play a large role and influence on his beliefs in African nationalism. Meanwhile, Mr. Delany got married and he and his wife ultimately raised seven children.

Mr. Delany also attended Jefferson College and worked in medicine under the tutelage of several prominent Pennsylvania doctors. In 1843, Mr. Delany began writing for a newspaper entitled, *The Mystery*, which was an abolitionist centric publication. It was during this time, that he developed the concept of establishing a 'Black Israel' on the east coast of Africa.

A rare portrait set of an Unidentified family. (LOC)

In 1847, Mr. Delany met with fellow abolitionist Frederick Douglass and William Lloyd Garrison and they decided to form a business relationship and publish an abolitionist newspaper entitled, *North Star*. Mr. Delany also continued his medical studies and in 1850, he won admission to Harvard Medical School. Ultimately, he was denied entry after whites protested the acceptance of he and three other blacks.

Consequently, Mr. Delany was frustrated with racial inequality and in 1852, he authored a book entitled, *The Condition, Elevation, Emigration, and Destiny of the Colored People of the United States, Politically Considered.* In his book, Mr. Delany listed the numerous successful free blacks who made great achievements because they had opportunity. He also highlighted the evils of servility, stating, *"With broken hopes-sad devastation; A race resigned to DEGRADATION!* In obvious frustration with pervasive racial inequality in the United States, he urged his fellow blacks to move to Africa as they would never be provided equal rights in the United States.

Mr. Delany then moved to Canada where he worked on several other abolitionist publications and helped with the northern leg of the Underground Railroad. In 1859, Mr. Delany sailed to Liberia in Africa where he traveled extensively for nine months. Mr. Delany forged bonds with several village chieftains to establish a new black nation but due to tribal differences, the plan failed. In 1860, Martin Delany embarked on a cause working for emancipation of slaves, and then in 1861, the American Civil War began. Mr. Delany returned to the United States.

In 1863, with President Lincoln's emancipation of slaves, Mr. Delany served as a recruiter and was instrumental in the establishment of several colored regiments. Mr. Delany joined the Army and served in the 52nd US Colored troops. He ultimately achieved the rank of Major, which was the highest rank held by any black in a combat unit. Recall that Dr. Alexander Augusta achieved the rank of Brevet Lieutenant Colonel, but he was a surgeon in the medical corps and did not serve in a line unit on the battlefield.

During the war, Major Delany, Frederick Douglass and others fought for pay equality for African American soldiers. White soldiers received $13.00 per month while blacks received $10.00. This disparity coupled with separate regiments and a predominantly white officer corps remained contentious issues throughout the war.

After the war, Martin Delany continued his fight for equal rights and he worked in several Reconstruction Era political initiatives focused on equality. In 1880, Mr. Delany and his family moved to Ohio, and in 1885, he died there of tuberculosis. Martin R. Delany was one of the hardest fighting yet seemingly

Major Martin R. Delany, The highest ranking African American line officer during the Civil War (NPG)

overlooked, unsung hero of African American causes.

The 107th United States Colored Troops Band. (LOC)

A proud pair of U.S.C.T. riflemen ready for battle. (LOC)

The Medal of Honor
as awarded during the Civil War (U.S. Army)

Chapter 5

Medal of Honor

"Rushed in advance of his brigade, shot a rebel officer who was on the parapet rallying his men, and then ran him through with his bayonet." – Private James Gardner (award citation).

Several other African Americans that wore the federal uniform in Union blue won the nation's highest award, the Medal of Honor. As Frederick Douglass said, *"Once let the black man get upon his person the brass letter, U.S., let him get an eagle on his button, and a musket on his shoulder and bullets in his pocket, there is no power on earth that can deny that he has earned the right to citizenship."* African Americans made fine military men and they served their nation with great distinction- they just simply needed an opportunity to prove themselves.

In fact, a total of twenty-five African Americans earned the Medal of Honor during the American Civil War. Eighteen soldiers and seven sailors comprise the ranks of the most valorous and highly decorated African Americans of the war.

The Battle of Mobile Bay, Alabama (LOC)

In the U.S. Navy, of the seven sailors that earned the Medal of Honor, four were won during the Battle of Mobile Bay. They and their fellow honored sailors along with their commendations are listed below:

Aaron Anderson, Navy Landsman of the USS Wyandank on March 17, 1865 in the battle of Mattox Creek, Virginia while "Participating with a boat crew in the clearing of Mattox Creek, L/man Anderson carried out his duties courageously in the face of a devastating fire which cut away half the oars, pierced the launch in many places and cut the barrel off a musket being fired at the enemy."

Robert Blake, Navy Contraband sailor of the USS Marblehead on December 25, 1863 fighting "an engagement with the enemy on John's Island [South Carolina]. Serving the rifle gun, Blake, an escaped slave, carried out his duties bravely throughout the engagement which resulted in the enemy's abandonment of positions, leaving a caisson and one gun behind."

William H. Brown, Navy Landsman of the USS Hartford during the Battle of Mobile Bay (August 5, 1864), whose citation reads, "Remained steadfast at his post and performed his duties in the powder division throughout the furious action which resulted in the surrender of the prize rebel ram (C.S.S. Tennessee) and in the damaging and destruction of batteries at Fort Morgan."

Wilson Brown, Navy Landsman of the USS Hartford during the Battle of Mobile Bay (August 5, 1864), whose citation reads, "Knocked unconscious into the hold of the ship when an enemy shell burst fatally wounded a man on the ladder above him, Brown, upon regaining consciousness, promptly returned to the shell whip on the berth deck and zealously continued to perform his duties although 4 of the 6 men at this station had been either killed or wounded by the enemy's terrific fire."

John Henry Lawson, Navy Landsman of the USS Hartford during the Battle of Mobile Bay (August 4, 1864), was cited for valor with, though "Wounded in the leg and thrown violently against the side of the ship when an enemy shell killed or wounded the 6-man crew as the shell whipped on the berth deck, Lawson, upon regaining his composure, promptly returned to his station and, although urged to go below for treatment, steadfastly continued his duties."

James H. Harris, of the 38th Colored Infantry Unit, who won the Medal of Honor for his actions in the Battle of Chaffin's Farm. (LOC)

Powhatan Beatty, First Sergeant, 5th U.S. Colored Infantry Regiment won the Medal of Honor for his actions in the Battle of Chaffin's Farm. (LOC)

Sergeant Major Christian Fleetwood, 4th U.S. Colored Infantry Regiment won the Medal of Honor for his actions in the Battle of Chaffin's Farm. (LOC)

Sergeant William H. Carney, 54th Massachusetts Colored Infantry Regiment, won the Medal of Honor for his actions at the Battle of Fort Wagner. (GC)

James Mifflin, Navy Cook of the USS Brooklyn during the Battle of Mobile Bay (August 4, 1864), "Remained steadfast at his post and performed his duties in the powder division throughout the furious action which resulted in the surrender of the prize rebel ram (C.S.S. Tennessee) and in the damaging and destruction of batteries at Fort Morgan."

Joachim Pease, Navy Seaman of the USS Kearsarge during a naval battle off the coast of Cherbourg, France on June 19, 1864, was commended for valor while "Acting as loader on the No. 2 gun during this bitter engagement, Pease exhibited marked coolness and good conduct and was highly recommended by the divisional officer for gallantry under fire."

Battlefield at Chaffin's Farm, Virginia - U.S.C.T. troops encamped upon the field. (LOC)

Fourteen African American USCT soldiers won the nation's highest award for valor at the bitterly contested Battle of Chaffin's Farm (also known as the Battle of New Market Heights). On September 29, 1864, while fighting against Confederates desperately defending the outskirts of Richmond, Virginia, an entire division of U.S. Colored Troops saw heavy combat. Among the men decorated with the Medal of Honor in this battle were:

William Barnes, Private, 38th U.S. Colored Infantry Regiment was cited with, "Among the first to enter the enemy's works; although wounded."

Powhatan Beatty, First Sergeant, 5th U.S. Colored Infantry Regiment "Took command of his company, all the officers having been killed or wounded, and gallantly led it."

James Bronson, First Sergeant, 5th U.S. Colored Infantry Regiment who "Took command of his company, all the officers having been killed or wounded, and gallantly led it."

Christian Fleetwood, Sergeant Major, 4th U.S. Colored Infantry Regiment was cited with, "Seized the colors, after 2 color bearers had been shot down, and bore them nobly through the fight."

James Gardner, Private, 36th United States Colored Infantry Regiment was commended with, "Rushed in advance of his brigade, shot a rebel officer who was on the parapet rallying his men, and then ran him through with his bayonet."

James H. Harris, Sergeant, 38th U.S. Colored Infantry Regiment was recognized for "Gallantry in the assault."

Thomas Hawkins, Sergeant Major, 6th U.S. Colored Infantry Regiment, cited for his "rescue of regimental colors."

Alfred Hilton, Sergeant, 4th U.S. Colored Infantry Regiment, was cited with, "When the regimental color bearer fell, this soldier seized the colors and carried it forward, together with the national standard, until disabled at the enemy's inner line."

Milton Holland, Sergeant Major, 5th U.S. Colored Infantry Regiment, was cited for, "When the regimental color bearer fell, this soldier seized the color and carried it forward, together with the national standard, until disabled at the enemy's inner line."

Miles James, Corporal, 36th U.S. Colored Infantry Regiment, who "Having had his arm mutilated, making immediate amputation necessary, he loaded and discharged his piece with one hand and urged his men forward; this within 30 yards of the enemy's works."

Alexander Kelly, First Sergeant, 6th U.S. Colored Infantry, who "Gallantly seized the colors, which had fallen near the enemy's lines of abatis (field obstacles), raised them and rallied the men at a time of confusion and in a place of the greatest danger."

Robert Pinn, First Sergeant, 5th U.S. Colored Infantry Regiment, who "Took command of his company after all the officers had been killed or wounded and gallantly led it in battle."

Edward Ratliff, First Sergeant, 38th U.S. Colored Infantry Regiment who "Commanded and gallantly led his company after the commanding officer had been killed; was the first enlisted man to enter the enemy's works."

Charles Veale, Private, 4th U.S. Colored Infantry Regiment, was cited commended with, "Seized the national colors after 2 color bearers had been shot down close to the enemy's works, and bore them through the remainder of the battle."

Battle of Chaffin's Farm, Virginia. Engraving of the U.S.C.T. troops in the assault.
(Harper's Weekly)

The remaining four USCT soldiers who won the Medal of Honor during the Civil War are comprised of various actions and units as listed below.

Bruce Anderson, Private, 142nd New York Volunteer Infantry, while fighting on January 15, 1865 at the 2nd Battle of Fort Fisher, North Carolina, he "Voluntarily advanced with the head of the column and cut down the palisading."

William H. Carney, Sergeant, 54th Massachusetts Colored Infantry Regiment, while fighting on July 18, 1863, at the Battle of Fort Wagner, South Carolina, "Grasped the flag, led the way to the parapet, and planted the colors thereon. When the troops fell back he brought off the flag, under a fierce fire in which he was twice severely wounded." The assault on Fort Wagner was graphically portrayed in the 1989 movie entitled, *Glory*.

Decatur Dorsey, Corporal, 39th United States Colored Infantry Regiment, during the Battle of the Crater at Petersburg, Virginia on July 30, 1864, he, "Planted his colors on the Confederate works in advance of his regiment, and when the regiment was driven back to the Union works he carried the colors there and bravely rallied the men." This action was depicted in the 2003, Hollywood movie entitled *Cold Mountain*.

Andrew J. Smith, Corporal, 55th Massachusetts Colored Infantry Regiment, while fighting on November 30, 1864 at the Battle of Honey Hill, South Carolina, he was cited for "Saving his regimental colors, after the color bearer was killed during a bloody charge called the Battle of Honey Hill, South Carolina."

Crew of the U.S.S. Miami. Note the U.S. Navy sailors are integrated. (NHC)

U.S.S. Hunchback. U.S. Navy sailors on the deck. Note the integrated crew. (NARA)

Unidentified U.S. Navy sailor with cigar and photos in hand. (LOC)

Crew of the ironclad warship, U.S.S. Monitor in 1862. The Monitor sank in heavy seas off the coast of Cape Hatteras, North Carolina on December 31, 1862. While 16 souls were lost at sea, 47 survived. The sole African American crewman in this photograph is 24- year-old, 5'6", Josiah "Siah" Carter. His rank was "1st Class Boy." According to his service file he survived the war and served on several other ships as well.

A sentry at his post in Virginia. (LOC)

Colonel Robert Gould Shaw, 54th Massachusetts Infantry Regiment (LOC)

4th United States Colored Infantry Regiment (U.S.C.T.) (LOC)

Close-up view of the same young men- note the pride and determination.

Lieutenant Peter Vogelsang and Sergeant Major John H. Wilson
of the 54th Massachusetts Colored Infantry Regiment (GC)

Chapter 6

Unknown Spirits

""Once let the black man get upon his person the brass letter, U.S., let him get an eagle on his button, and a musket on his shoulder and bullets in his pocket, there is no power on earth that can deny that he has earned the right to citizenship."

- Frederick Douglass

     As one studies the Civil War era, photographs provide invaluable insights to life during those times. The transition from life as a slave to that of a free man would have been nothing less than a tremendous upheaval of the soul. The depth of such a life-changing event could only truly be known by someone who experienced such a transformation. It is extremely rare to find photographs of former slaves who were so fortunate as to become free men and then transitioned into soldiers. However, there are several examples of these rare photographically documented transitions.

Hubbard Pryor-from contraband to soldier (NARA)

One example can be found in that of former slave Hubbard Pryor. In 1864, he transformed from contraband to proud soldier of the 44th Colored Infantry Regiment (U.S.C.T.). Another set of similar transformations is that of "Gordon." Recall from a previous photograph, the former slave as captured on film with the horrific keloid scars displayed on his back. "Gordon," somehow found freedom from the oppressive bondage he once endured on a Louisiana plantation, and he became a soldier too. His story was told in a *Harper's Weekly* periodical published on July 4, 1863. However, the story only reports his name as "Gordon," "A Typical Negro." During the Civil War, abolitionists distributed the cabinet card displaying the image of "Gordon" as well.

There is however a little more information that can be pieced together regarding his story. An article was printed in another Civil War newspaper from Boston, entitled *The Liberator* on June 12, 1863 that reports a little more information. The article states:

> *There has lately come to us, from Baton Rouge, the photograph of a former slave—now, thanks to the Union army, a freeman. It represents him in a sitting posture, his stalwart body bared to the waist, his fine head and intelligent face in profile, his left arm bent, resting upon his hip, and his naked back exposed to full view. Upon that back, horrible to contemplate! is a testimony against slavery more eloquent than any words. Scarred, gouged, gathered in great ridges, knotted, furrowed, the poor tortured flesh stands out a hideous record of the slave-driver's lash. Months have elapsed since the martyrdom was undergone, and the wounds have healed, but as long as the flesh lasts will this fearful impress remain. It is a touching picture, an appeal so mute and powerful that none but hardened natures can look upon it unmoved. However, much men may depict false images, the sun will not lie. From such evidence as this there is no escape, and to see is to believe. Many, therefore, desired a copy of the photograph, and from the original numerous copies have been taken.*
>
> *The surgeon of the First Louisiana regiment, (colored,) writing to his brother in the city, encloses this photograph, with these words: –*
>
> *"I send you the picture of a slave as he appears after a whipping. I have seen, during the period I have been inspecting men for my own and other regiments, hundreds of such sights—so they are not new to me; but it may be new to you. If you know of anyone who talks about the humane manner in which the slaves are treated, please show them this picture. It is a lecture in itself."*

Yet another article was run by the same paper a week later, entitled "*The Scourged Back.*" This article states:

> *We received from Baton Rouge the photographic likeness of a slave's naked back, lacerated by the whip [...] We look on the picture with amazement that cannot find words for utterance. Amazement at the cruelty which could perpetrate such an outrage as this; at the brutal folly, the stupid ignorance, that could permit such a piece of infatuation; at the absence, not only of humane feeling, but of economical prudence of common sense, of ordinary intelligence, displayed in such frantic thoughtlessness. Among what sort of people are such things possible? [...] This card-photograph should be multiplied by the hundred thousand, and scattered over the states. It tells the story in a way that even Mrs. Stowe cannot approach; because it*

tells the story to the eye. If seeing is believing — and it is in the immense majority of cases — seeing this card would be equivalent to believing things of the slave states which Northern men and women would move heaven and earth to abolish!

These articles pieced together with references to the image accompanying "Gordon" later as a soldier refer to the man in the image as "Whipped Peter," of the Corps d' Afrique. This unit was comprised of U.S. Louisiana Colored troops. A search of the archival records of the National Archives indicates there was a man named Peter Gordon who served in the 73rd Regiment (U.S.C.T.). The 1st Regiment of the Corps d' Afrique was organized from the 1st Louisiana Native Guards, which later became the 73rd U.S.C.T.

"Gordon's" escape from slavery and his arrival in camp all fit within a plausible time frame. While we in the modern world may never know for certain who exactly the man known as "Gordon" really was, the images coupled with the few threads of information we do have weaves a fascinating mystery.

"Gordon" or "Whipped Peter"- Composite images.
(Harpers Weekly)

The remaining imagery in this chapter is devoted to people who are tragically listed in historic photographic collections as "unknown" or "unidentified" African American. The bulk of these photographs are from the Liljenquist and Gladstone collections housed in the Library of Congress. There will be no additional narrative in this chapter, just the exquisite yet haunting imagery of unknown spirits of the past.

Soldiers and Sailors...

Women and children too...

Transitioning from Contraband to Soldier-Unknown **(LOC)**

Unknown colored troops digging Dutch Gap Canal, Virginia **(LOC)**

Integrated troops… but all equally… unknown.

Chapter 7

Courage and Pride

"Boys, I only did my duty; the old flag never touched the ground." - Sergeant William Carney

The Medal of Honor has an interesting history that began during the American Civil War. At the beginning of the war, there was no medal for valor in the U.S. military but there was a "Certificate of Merit" that had been in place since the Mexican-American War (1846-1848). In 1861, a Navy Medal of Valor was established by Congress via Bill 82: 37th Congress, Second Session, 12 Statute 329, "to promote the efficiency of the Navy" and "to be bestowed upon such petty officers, seamen, landsmen, and marines as shall most distinguish themselves by their gallantry and other seamen-like qualities during the present war."

A courageous and proud young Corporal- ready for action. (LOC)

Then, in 1862, Congress established a similar "Medal of Honor" for the U.S. Army. The Army medal was the equivalent of the Navy medal and its purpose was to recognize "non-commissioned officers and privates as shall most distinguish themselves by their gallantry in action and other soldier-like qualities during the present insurrection," 37th Congress, Second Session, 12 Statute 623. The Secretary of War had the medal's purpose altered to be more comprehensive with, "to provide for the presentation of medals of honor to the enlisted men of the army and volunteer forces who have distinguished or may distinguish themselves in battle during the present rebellion."

The War Department contracted with William Wilson and Son, a silversmith firm in Philadelphia for the medal's design. The U.S. Mint in Philadelphia was then directed to stamp out the medals from a mixture of copper and bronze. On the back of the Navy award, the medal states "Personal Valor," and the reverse of the Army version states, "The Congress to." When prepared for presentation, the medal was engraved with the name of the recipient.

In 1863, the Medal of Honor was made a permanent award, not just for the rebellion, and eligibility for the award was extended to officers as well as enlisted. Today, the award remains the nation's highest and most prestigious honor in all the military branches. While the award has undergone several physical changes since the Civil War, there have also been changes regarding eligibility and criteria for the award.

During the Civil War, in principle, there were no exclusions for black soldiers, and the first Medal of Honor bestowed upon an African American soldier went to a former slave from Virginia, Sergeant William H. Carney. As stated by Sergeant Carney, *"Previous to the formation of colored troops, I had a strong inclination to prepare myself for the ministry; but when the country called for all persons, I could best serve my God serving my country and my oppressed brothers."* The *Liberator*, November 6, 1863.

Sergeant William H. Carney (1840-1908) was awarded his medal for heroic actions during the Battle of Fort Wagner, South Carolina. In that action, Sergeant Carney served as a color bearer for the United States Army. Carrying the flag has always been an honorable and important military task but during the Civil War, the flags served as a symbol of both pride and battlefield communications.

Sergeant William H. Carney, Color Bearer 54th Massachusetts, Medal of Honor. (LOC)

Confederate defenses south of Charleston, South Carolina. 1862 Union map. (LOC)

Other troops looked to the flags for field positions, maneuvering formations, rally points and very importantly, inspiration. Consequently, the enemy also focuses on eliminating the flag bearers of the opposing force. At dusk on July 18, 1863 the 54th Massachusetts charged through deep sands attacking a huge fortification defended with numerous cannons and mortars. In addition to the artillery fire, 1,600 Confederates poured musket fire into the Union attackers from reinforced parapets above them. While the 54th Massachusetts was supported by several other Union regiments, the battleground before the fort was narrow and restricted the assaulting forces.

The 54th Massachusetts storming Confederate Fort Wagner. Colonel Robert Shaw is depicted alongside Sergeant William Carney. (LOC)

Sergeant Carney was shot four times, twice in his torso, his arm and head, yet he carried the flag atop the fort. He then kept it flying for at least twenty minutes, serving as an inspiration to his fellow men. Against such overwhelming odds, ultimately, the Union troops were driven back by the Confederate defenders. During the retreat from the fort, Sergeant Carney stumbled back to safety and passed the colors to another man, saying, *"Boys, I only did my duty; the old flag never touched the ground."* Miraculously, the twenty-three-year old Sergeant Carney never dropped the flag and he survived his wounds.

At the Battle of Fort Wagner, the fighting was some of the bloodiest and intensive slaughter of the war. The battle pitted South Carolina whites in the Confederacy (who held defensive positions in the fort) against an assaulting Union force led by the 54th Massachusetts Colored troops.

The battle was deeply personal for both sides but especially for the African Americans, many of whom were former slaves returning to fight as U.S. Army soldiers.

Within the ranks of the 54th Massachusetts were also men such as Sojourner Truth's grandson, Private James Caldwell and Frederick Douglass's son, Sergeant Major Lewis Douglass. More importantly, the pride of many people marched with those young men into battle on that fateful day.

JEREMIAH ROLLS, 1st Sergt., Co. I.　　　ABRAM C. SIMMS, CORP., Co. I.
　　GEORGE LIPSCOMB, CORP., Co. I.　THOMAS BOWMAN, Sergt., Co. I.
ISOM AMPEY, Pvt., Co. K.　　　　　　JOHN H. WILSON, Sergt. Major.

Six members of the 54th Massachusetts Infantry Regiment, U.S.C.T.
(LOC)

During the Civil War, white officers in the USCT led the black enlisted men into battle. Colonel Robert Gould Shaw, a young twenty-five-year-old man from Massachusetts led the charge and he gave his life for his men and his country. Robert Shaw was a recent graduate of Harvard University and had only been married for one month before he died. His widow never remarried. The events of the Battle of Fort Wagner are graphically depicted in the 1989 movie entitled, *Glory*.

Fort Wagner: The earthen berms, bombproofs and obstacles from two perspectives. The Confederate stronghold was never taken in battle, it was abandoned as the Confederates retreated from Morris Island. (LOC)

One important lesson learned from the Civil War is no matter what color a man's skin may be, we are all equal in God's eyes and we all bleed red blood. The shifting sands of Morris Island are forever stained with the blood of many men, black and white. The 54th Massachusetts attacked Fort Wagner with nearly 600 men and they suffered nearly 50% casualties.

Fort Wagner: Interior view. The earthen berms, bombproofs and obstacles from two perspectives (LOC).

With the stench of hundreds of dead bodies swelling in the July sun, the men had to be buried. The Confederates buried Colonel Shaw and his men in a mass grave in the sand dunes outside the fort. After the war, the U.S. government had the bodies disinterred and reburied nearby at Beaufort National Cemetery. Fort Wagner is long gone, destroyed by waves and shifting sands for more than one-hundred and fifty years of natural cleansing. Today the island is a desolate, dune filled natural preserve and marked only by the peaceful sounds of the sea.

Post-battle casualty collection. African Americans tasked with retrieving remains from the battlefield. (LOC)

Memorial bronze honoring the courage and pride of the 54th Massachusetts Infantry.

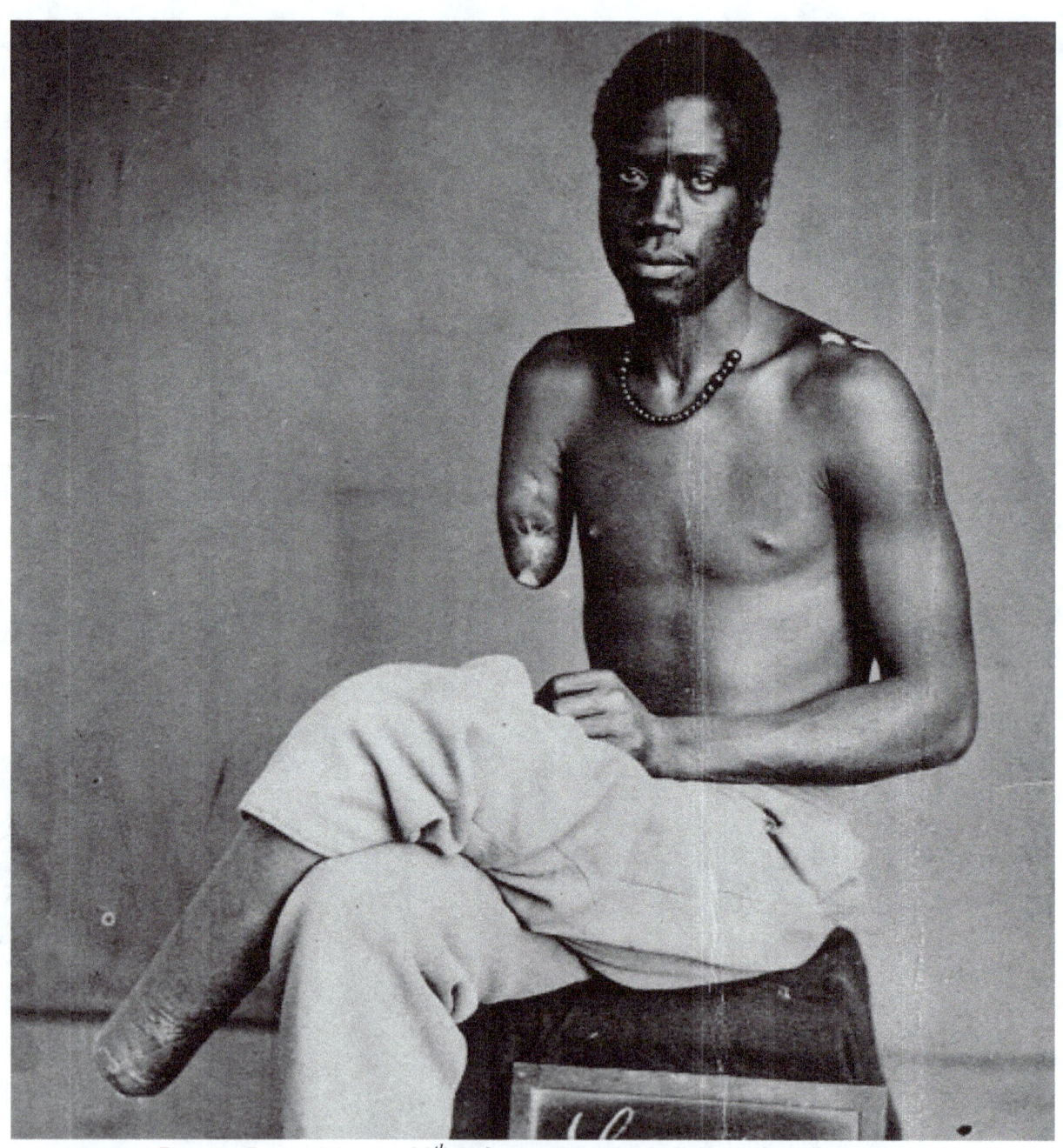

Private Lewis Martin, 29th *Infantry Regiment, U.S. Colored Troops. Double amputee, but he survived the war and died in 1892.* (NARA)

Chapter 8

The Nurses

I have seen the terrors of that war. I was the wife of one of those men... - Susan Baker King Taylor

      While hundreds of thousands of men died on the battlefields and in the diseased encampments of the Civil War, many survived thanks to the selfless work of the men and women of the medical profession. A soldier's life in the field is full of suffering and degradation. Often the men marched without shoes, amongst the elements and many received no pay for extremely long lengths of time. These soldiers were totally dependent on the government for their care.

      When the men got sick, medical care was limited and when they were badly wounded, they were primarily in the hands of God, as there was not much man could do for them. Amongst the unsung heroines of the war, nurses toiled under harsh conditions to render aid and save lives. One of these "Angels of Mercy" caregivers was Nurse Susie King Taylor who worked in the blood-stained sands of coastal South Carolina.

      Susan Baker King Taylor (1848-1912) was born a slave in Liberty County, Georgia. As a young girl, she was moved to Savannah where she was secretly taught to read and write by other slaves and compassionate whites. As Susie matured, she honed her skills in literacy and she became a teacher as well.

      When the Civil War erupted in 1861, Miss Baker's master moved her back to the rural countryside as Union soldiers secured nearby Fort Pulaski and violence was imminent along coastal Georgia and South Carolina. On April 1, 1862, fourteen-year-old Miss Baker and other members of her family fled to Union lines. Upon arrival at the Union garrison on St. Simon's Island, she and her family won their freedom. The Union army put the free but poverty stricken blacks to work and Commodore Louis M. Goldsborough offered Miss Baker books and supplies if she would organize a school.

In short order, Miss Susie Baker was teaching children, and in the evenings, she taught adults. In fact, by accepting her assignment, Miss Baker also became the first African American teacher in the State of Georgia. One of her students was a bright young soldier named Edward King. Susie and Sergeant King of Company E, 1st South Carolina Volunteers, were married and she moved along with his regiment over the next three years of the war.

School for colored folk near Beaufort, South Carolina (LOC)

Her duties varied from teacher to laundress and nurse. Her nursing duties began to predominant all other tasks as the war raged and diseases ravaged through the camps. Her husband's unit was the first federally authorized black regiment, the 1st South Carolina Infantry Regiment Volunteers but they were changed to the 33rd U.S. Colored Troops. These soldiers fought in the coastal regions and sea islands of Georgia and South Carolina. Mrs. King's duties were difficult work under extremely harsh conditions especially during the sweltering heat of the summer.

1st South Carolina Colored Infantry in formation near Beaufort, South Carolina (LOC)

The siege warfare in and around Charleston, South Carolina was long and extremely violent. All too soon, Sergeant King's unit was sent into a series of fighting and several major engagements with the enemy. Saying goodbye to her husband was very difficult and in the words of Mrs. Susie King,

I went with them as far as the landing, and watched them until they got out of sight, and then I returned to the camp. There was no one at camp but those left on picket and a few disabled soldiers, and one woman, a friend of mine, Mary Shaw, and it was lonesome and sad, now that the boys were gone, some never to return.

Mary Shaw shared my tent that night, and we went to bed, but not to sleep, for the fleas nearly ate us alive. We caught a few, but it did seem, now that the men were gone, that every flea in camp had located my tent, and caused us to vacate. Sleep being out of the question, we sat up the remainder of the night.

While assigned to Camp Shaw and working at the nearby Union military hospital at Beaufort, South Carolina, Mrs. King met Ms. Clara Barton, known as the "Angel of Mercy" who later founded the American Red Cross. Mrs. King recorded in her autobiography, written after the war, that Ms. Barton was "always very cordial toward me, and I honored her for her devotion and care of those men."

Miss Clara Barton, Founder of the American Red Cross. (LOC)

Mrs. King weathered many long days and tortuous nights during the last few years of the war. While Susie King remained in the unit's main encampment on the sandy shores of Folly Island, South Carolina, her husband's unit fought nearby in the battles of Honey Hill, James Island and ultimately the capture of Charleston. While the 33rd USCT suffered many casualties, Sergeant King survived and he and Susie were reunited at the war's end.

Union camp at Folly Beach, *South Carolina* (LOC)

For the survivors, the Union destruction of the Confederate States of America was a joyous yet melancholy episode in history. After four long years of bloody fighting, the nation was deeply wounded but peace did bring glorious rewards to the former slaves.

U.S Colored Troops in fortifications outside Charleston. (LOC)

When the 33rd U.S.C.T. was mustered out at Morris Island, South Carolina, one of the senior officers, Lieutenant Colonel C.T. Trowbridge, stated the following to his men:

Soldiers, you have done your duty and acquitted yourselves like men who, actuated by such ennobling motives, could not fail; and as the result of your fidelity and obedience you have won your freedom, and oh, how great the reward! It seems fitting to me that the last hours of our existence as a regiment should be passed amidst the unmarked graves of your comrades, at Fort Wagner. Near you rest the bones of Colonel Shaw, buried by an enemy's hand in the same grave with his black soldiers who fell at his side; where in the future your children's children will come on pilgrimages to do homage to the ashes of those who fell in this glorious struggle.

At war's end, Mrs. Susie King was filled with emotions and she recorded them with:

What a wonderful revolution! In 1861 the Southern papers were full of advertisements for "slaves," but now, despite all the hindrances and "race problems," my people are striving to attain the full standard of all other races born free in the sight of God, and in a number of instances have succeeded. Justice we ask, to be citizens of these United States, where so many of our people have shed their blood with their white comrades, that the stars and stripes should never be polluted...

We do not, as the black race, properly appreciate the old veterans, white or black, as we ought to. I know what they went through, especially those black men, for the Confederates had no mercy on them; neither did they show any toward the white Union soldiers. I have seen the terrors of that war. I was the wife of one of those men who did not get a penny for eighteen months for their services, only their rations and clothing.

After the war, Mr. and Mrs. King returned to Savannah and Liberty County, Georgia where Susie established several schools and once again she followed her passion, teaching school. Edward became a carpenter and laborer as there was a lot of necessary rebuilding following the war.

In 1866, just before the birth of their only child, Mrs. King's husband Edward suddenly died. As a distraught and lonely widow, Mrs. King struggled to maintain her schools and raise her child, but she could not do both all on her own. In 1872, she went to work as a laundress for a wealthy white lady and they moved to Boston, Massachusetts.

While living in Boston, Susan King married Mr. Russell Taylor and she remained there for the rest of her life. However, before she died, in 1902, she published her autobiography entitled, *Reminiscences of My Life in Camp*, which is a treasure for the modern world.

U.S. Colored Troops in camp. (LOC)

Her rare insights to life as a slave and her exploits during the Civil War record the emotional experiences of a very intelligent, strong willed yet compassionate African American lady. Finally, in her own words, Mrs. Susan Baker King Taylor wrote the following passage in her book,

> *I look around now and see the comforts that our younger generation enjoy, and think of the blood that was shed to make these comforts possible for them, and see how little some of them appreciate the old soldiers. My heart burns within me, at this want of appreciation. There are only a few of them left now, so let us all, as the ranks close, take a deeper interest in them. Let the younger generation take an interest also, and remember that it was through the efforts of these veterans that they and we older ones enjoy our liberty to-day.*

Susan Baker King Taylor dedicated her autobiography to the commander of the 33rd U.S.C.T., Union officer, Colonel Thomas Wentworth Higginson. Susan Baker King Taylor died in 1912 and her book provides a unique and insightful view of the Civil War.

Susan Baker King Taylor, U.S.C.T. Nurse. As depicted in
*Reminiscences of My Life in Camp with the
33D United States Colored Troops Late 1st S.C. Volunteers.*

While the Civil War was a long and bloody episode for the nation, for African Americans, it brought the dawn of a new day and new beginnings. During the Civil War, nearly 180,000 men, 10% of the Union Army, served in the U.S. Army and nearly 20,000 served in the Navy. Roughly 40,000 African American soldiers died in the war, mostly due to disease. If not for the selfless service of women such as Mrs. Susan Baker King Taylor and so many others whose names are lost to history, the casualty figures would have been even higher.

Another member of the same unit Mrs. Taylor served in, the 1st South Carolina Colored Volunteers was Sergeant Prince Rivers. The history of Mr. Rivers is difficult to trace, but in local South Carolina, his exploits are somewhat legendary. As with many former slaves, their histories lack a solid foundation in historical fact due to the lack of available records. However, there are several facts and primary sources that support elements of his story. When pieced together, the legend of Prince Rivers is quite remarkable.

Mr. Rivers was born a slave in 1822 on a plantation in Beaufort, South Carolina. His master was Henry Middleton Stuart, Sr. of "Oak Point" plantation. As an adult slave, Mr. Rivers served on the household staff and as carriage driver, an esteemed position for a slave. He also learned to read and write and according to military records, he was "a stalwart man, fully six feet in height, broad shouldered and robust, complexion black."

In 1862, Mr. Rivers stole his master's horse and rode to safety within the Union lines outside Beaufort. Mr. Rivers then enlisted in the U.S. Army, which was then forming the first colored regiment. Mr. Rivers was assigned to Hunter's Brigade of the 1st South Carolina Infantry Volunteers. This unit was later re-designated the 33rd U.S. Colored Troops regiment. Mr. Rivers was forty years old and quickly identified by the white leadership for his ability to lead. Promoted to the rank of Sergeant, Prince Rivers was also assigned to multiple leadership roles.

As recorded by his commanding officer, Colonel Thomas Wentworth Higginson, Sergeant Rivers was a respected soldier. His book states:

> *Sergeant Prince Rivers, our color-sergeant, who is provost-sergeant also, and has entire charge of the prisoners and of the daily policing of the camp. He is a man of distinguished appearance, and in old times was the crack coachman of Beaufort, in which capacity he once drove Beauregard from this plantation to Charleston, I believe. They tell me that he was once allowed to present a petition to the Governor of South Carolina in behalf of slaves, for the redress of certain grievances; and that a placard, offering two thousand dollars for his recapture, is still to be seen by the wayside between here and Charleston.*
>
> *He was a sergeant in the old "Hunter Regiment," and was taken by General Hunter to New York last spring, where the chevrons on his arm brought a mob upon him in Broadway, whom he kept off till the police interfered. There is not a white officer in this regiment who has more administrative ability, or more absolute authority over the men; they do not love him, but his mere presence has controlling power over them. He writes well enough to prepare for me a daily report of his duties in the camp; if his education reached a higher point, I see no reason why he should not command the Army of the Potomac.*
>
> *He is jet-black, or rather, I should say, wine-black; his complexion, like that of others of my darkest men, having a sort of rich, clear depth, without a trace of sootiness, and to my eye very handsome. His features are tolerably regular, and full of command, and his figure*

superior to that of any of our white officers,—being six feet high, perfectly proportioned, and of apparently inexhaustible strength and activity. His gait is like a panther's; I never saw such a tread. No anti-slavery novel has described a man of such marked ability. He makes Toussaint perfectly intelligible; and if there should ever be a black monarchy in South Carolina, he will be its king.

Sergeant River's unit served in the occupation of Jacksonville, Florida and they fought at the Battle of Honey Hill and the capture of Confederate forces at James Island. Following garrison duties on the shores southeast of Charleston, they were mustered out of service at Fort Wagner, South Carolina on February 9, 1866.

After the war, Prince Rivers moved to a farm on the outskirts of Hamburg, South Carolina. In 1867, he was elected to the State Constitutional convention and he represented Edgefield County in the State Legislature for several years. He was also appointed to the position of Major-General in the South Carolina National Guard by Governor Robert Scott. He also served as a Chairman of the Republican party representing the newly established, Aiken County. He then served for three years as the Mayor of the town of Hamburg and several other civic positions, county coroner and justice of the peace.

Emancipation Day celebration- Sergeant Prince Rivers addressing his fellow troops.
(Frank Leslie's Illustrated Newspaper, January 24, 1863)

During his tenure as Mayor of Hamburg, he witnessed the infamous riots of Hamburg in 1876. These riots were the result of Reconstruction era racial friction and conflict. On July 4, 1876, during an Independence Day parade, the local black Reconstruction militia was confronted by an angry mob of white protesters. The heated confrontation became violent and Mayor Rivers intervened. While the mayor successfully quelled the altercation, the white men gave an ultimatum for the blacks to surrender their weapons and disband the militia by the 8th of July.

On the evening of July 8, 1876, a white mob returned to the headquarters of the local black militia. Armed with firearms, the white mob numbering at least one-hundred men secured the local militia headquarters and they took the black militiamen as prisoners. During the night, the white mob released the bulk of the black prisoners, but at least six black men were executed by the whites. Violence spread throughout South Carolina into a serious post-war racial conflict.

Following the murders, the white men proudly and publicly displayed the bloody shirts of their victims. The whites wore dyed red shirts which they adopted as the symbol of the "Red Shirt Movement." During the summer of 1876, throughout South Carolina, the movement gained in popularity amongst the white population. When state elections were held that fall, the "Red Shirts" intimidated blacks and stuffed the ballot boxes. Their goals were to eliminate Republican rule and intimidate blacks. Former Confederate General and Democratic gubernatorial candidate, Wade Hampton was elected as the State Governor. Consequently, while ninety-seven white men were named as the Hamburg murderers, justice was contravened by cronyism, and no one was prosecuted. Mayor Rivers was also terrorized, his property stolen and his farm was destroyed.

The U.S. Senate held hearings about the racial violence in South Carolina but it changed nothing in the South. While federal troops occupied the Southern states during Reconstruction, they could not effectively police the entire region. President Rutherford B. Hayes ordered the removal of the Union Army from the state of South Carolina on April 3, 1877. During this era, the Ku Klux Klan (KKK) was also organized and they terrorized African Americans as well. Consequently, the KKK, Red Shirts and other white supremacy groups perpetrated acts of intimidation, violence and outright terrorism against blacks throughout the South.

The "Red Shirts" succeeded in ending Republican rule for many years in the South. They also outlasted federal intervention, and violence against blacks continued into the 20th century. The Red Shirts movement was active in other southern states as well. Historians estimate that in addition to harassment and voting law violations in South Carolina, nearly one-hundred and fifty African Americans were killed by the Red Shirts during Reconstruction.

Following the riots, murders, and the fraudulent elections throughout the State, Mr. Rivers retired to Aiken, South Carolina. During his life, Mr. Rivers was twice married and had two children. In his elder years, Mr. Rivers suffered from frequent hemorrhages, and ultimately, he died of Bright's disease on April 10, 1887.

Per the obituary of Prince Rivers,

"In point of intelligence he was far above the average darkey, and was respected by the white people and looked up to by his colored brethren with much reverence. For a number of years he has been a trusted employee of Mr. B.P. Chatfield at the Highland Park Hotel. Attired in his livery suit and tall beaver hat, as a driver for Mr. Wm. C. Langley, he looked like a piece of statuary, so erect was he." His funeral was held at Cumberland A.M.E. Church, and he was buried in the Randolph Cemetery, Columbia, South Carolina."

- Aiken Review, 13 April, 1887

Racial division during Reconstruction. (Harper's Weekly)

Recently liberated slaves on a South Carolina plantation. (LOC)

Chapter 9

Mrs. Tubman

Master Lincoln, he's a great man, and I am a poor negro; but the negro can tell master Lincoln how to save the money and the young men. – Mrs. Harriet Tubman

Another African American lady that served the Union as a nurse was also an armed scout and spy. Born into slavery in Maryland as Araminta "Minty" Ross, she later changed her name to Harriet Tubman (1822-1913) to obscure both her marriage and her exploits for her own protection. Mrs. Tubman is one of the most famous abolitionist of the nineteenth century and her selfless services in the Underground Railroad have secured her name and reputation for all time.

As a young slave, Miss Ross (Harriet Tubman) was beaten severely by her master and she suffered a horrible head wound that caused her dizziness and pain for the rest of her life. She was also a devoted Christian and she despised slavery, oppression and inequality. In 1844, Miss Ross met and married a free black named John Tubman. In addition to changing her surname to Tubman, she also adopted a new first name, Harriet. These measures were prudent as she needed to conceal her identity for her own health and safety.

In 1849, Mrs. Tubman fled from bondage and traveled to Philadelphia, Pennsylvania but she soon returned and helped her family escape as well. Mrs. Tubman's husband chose to remain in Maryland and he abandoned her. Having secured a safe route to freedom, she returned and aided several groups of slaves to escape.

Mrs. Tubman became known as "Moses" amongst her people and she liberated dozens upon dozens of slaves. She then helped her people find jobs along with their newfound freedom. Mrs. Tubman developed routes and secured secret alliances as well as safe locations for her passengers to hide.

Mrs. Harriet Tubman, Conductor of the Underground Railroad, Union Scout and Spy (LOC)

The Underground Railroad by *Charles T. Webber.* (Cincinnati Art Museum)

In 1861, with the outbreak of war, Mrs. Tubman joined the Union Army where she worked as a cook and nurse in and around Port Royal, South Carolina. Her first battles of the war were against "camp fevers" such as dysentery and smallpox. Using local herbs and plants to treat her patients, she was apparently blessed by God as she never suffered from the illnesses she treated.

Contrabands on their journey northward- Union soldiers behind them are a good sign they made it to safety. (LOC)

Early in the war, Mrs. Tubman praised President Lincoln but she wanted him to free all the slaves and let them fight. In her own words, she reflected on her advice for the president:

God won't let master Lincoln beat the South till he does the right thing... Master Lincoln, he's a great man, and I am a poor negro; but the negro can tell master Lincoln how to save the money and the young men. He can do it by setting the negro free. Suppose that was an awful big snake down there, on the floor. He bite you. Folks all scared, because you die. You send for a doctor to cut the bite; but the snake, he rolled up there, and while the doctor doing it, he bite you again. The doctor dug out that bite; but while the doctor doing it, the snake, he spring up and bite you again; so he keep doing it, till you kill him. That's what master Lincoln ought to know.

Mrs. Tubman's straight forward approach to life soon saw her working as a volunteer scout for the Union. Travelling throughout the sandy shores and marshes of South Carolina and Florida, she used her covert skills learned during her tenure in the Underground Railroad and spied on the Confederate military. She also drew maps and provided information directly to Colonel James Montgomery of the US Army.

On one of her daring missions, she guided three steamships through mine-infested waters on a raid. On June 2, 1863, the successful Combahee River Raid resulted in the liberation of 750 slaves. Many of the men joined the U.S. Army Colored Troops. Harriet Tubman then began working as a recruiter. The press also picked up her story and reported her many successful operations of her service to the nation.

Combahee River Raid, **Harriet Tubman and the liberation of 750 slaves.**
(Harper's Weekly, July 4, 1863)

Mrs. Tubman served throughout the war and when it was over, she moved to New York. She also got married again, this time to a veteran of the war, Nelson Davis. She worked diligently on women's rights issues and was quite vocal regarding her concerns. In her elderly years, she joined the women suffrage movement and she was close friends with Susan B. Anthony and Emily Howland. It wasn't until the 1890s before the government added her to the military veteran pension rolls and she had to fight for any official recognition.

African American homecoming at war's end. (LOC)

Harriet Tubman- Post-war friends and family in Auburn, New York, circa 1887. Left to right: Harriet Tubman; Gertie Davis (Watson), adopted daughter; Nelson Davis, husband and veteran of the 8th USCT; Lee Chaney, neighbor's child; "Pop" John Alexander, elderly boarder in Tubman's home; Walter Green, neighbor's child; Blind "Aunty" Sarah Parker, elderly boarder; Dora Stewart, great-niece and granddaughter of Tubman's brother, Robert Ross.
Courtesy of The New York Times photo archive.

With the arrival of the 20th century, Mrs. Tubman was also a member of several other key organizations and charities. Among her work and affiliations, she was a member of the African Methodist Episcopal Zion Church and the National Federation of Afro-American Women. She donated freely to charities and she provided the real estate for the construction of a rest home for "aged and indigent colored people." Sadly, due to frailty she was forced to live in the facility she worked so hard to build for others. Harriet Tubman died surrounded by her family and friends in 1913 and she is buried in Fort Hill Cemetery, Auburn, New York.

Today, Mrs. Harriet Tubman and all the African Americans herein are considered heroic national treasures and some of the most influential citizens in U.S. history. In December 2014, the United States Congress designated a National Park to Harriet Tubman with one established on the Eastern Shore of Maryland, and another is in Auburn, New York which includes her residence, Home for the Aged, her church, Thompson AME Zion Church, and her grave site as well.

Due to nineteenth century laws regarding women in the military, ladies such as Harriet Tubman and Susan King Taylor only received partial compensation for their service. Ultimately upon appeal, they did receive a delayed government pension, years after the war. For many other women who supported the Union military, they never received any acknowledgement at all regarding their sacrifice and service. This was due to the loss of records during the war and sadly, blatant inequality.

While equitable compensation and continued inequality remained serious issues, the war did provide many opportunities for African Americans. Within five years of the war's end, the first African American in history was elected to the United States Congress. In 1870, Hiram Revels of Mississippi, was elevated from the state legislature to that of the U.S. Senate.

Hiram R. Revels (1827-1901) was born in 1827, in Fayetteville, North Carolina. His parents were free blacks and revels was tutored as young child and later attended institutions of Seminary schools in Indiana and Ohio. In 1845, Revels was ordained as a Minister in the African Methodist Episcopal Church (AME). From 1845- 1855, Minister Revels preached in churches throughout the Midwest.

During this time, Hiram Revels married a free black woman from Ohio, Ms. Phoebe A. Bass. The couple had six daughters. From 1855-1857, he went back to Seminary school where he had advanced religious studies. In 1858, he joined the Methodist Episcopal Church and he preached in Baltimore, Maryland. He also served as a local high school principal.

In 1861, Hiram Revels helped with U.S. Army recruiting and he helped raise several Colored regiments. In 1863, Mr. Revels joined the U.S. Army too, as a Chaplain. Chaplain Revels deployed to Mississippi and served the spiritual needs of soldiers during the Siege of Vicksburg.

After the war, Minister Revels served the AME church and in 1866, he was sent to Natchez, Mississippi where he and his family settled into a comfortable community. In 1868, upon urges from the local community, Minister Revels won a seat in the local government as an Alderman. In 1869, Revels won another political seat, this time as a State Senator representing Adams County. Then, in 1870, Revels was elected to the U.S. Senate for the State of Mississippi.

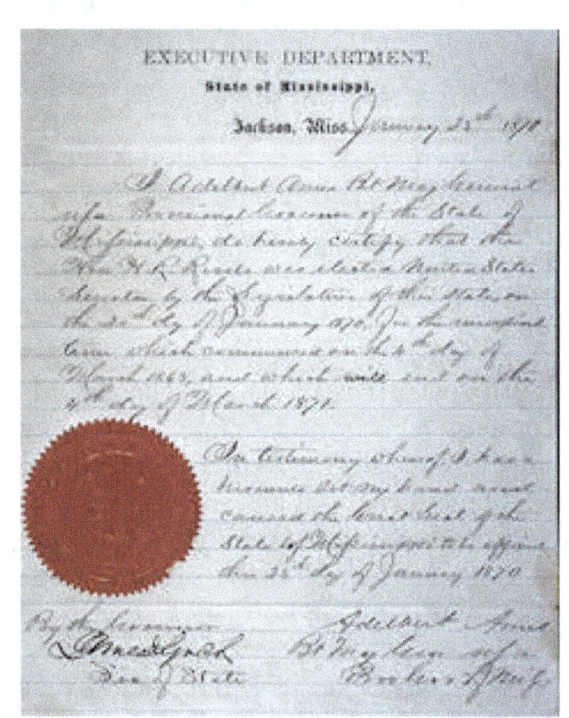

Certification letter of Senator Hiram Revels.
Dated January 25, 1870.
(State of Mississippi)

Senator Revels was highly respected and he always sought equality in legislative decisions. The newspapers praised the senator for his articulate skills as an orator. Although the black community generally disagreed with his position on amnesty for ex-Confederates, Senator Revels honorably completed his term in office, and in 1872, he returned to Mississippi. Back at home, citizen Revels served his community again as the co-founder of Alcorn University. He served there until 1873 as the school's first president. Hiram Revels returned to serve his state in an appointment capacity as Mississippi's Secretary of State.

By 1875, Hiram Revels was disgusted by politics and especially the northern carpetbaggers who had descended into the Southern States under the pretense of helping blacks to understand politics. Their schemes were clear to Mr. Revels. In a letter dated November 6, 1875, addressed to the President of the United States, Ulysses S. Grant, Hiram Revels expressed his concerns with:

> *Since Reconstruction, the masses of my people have been, as it were, enslaved in mind by unprincipled adventurers, who, caring nothing for country, were willing to stoop to anything no matter how infamous, to secure power to themselves and perpetuate it. My people are naturally Republicans and always will be, but as they grow older in freedom so do they in wisdom. A great portion of them have learned that they were being used as mere tools, and, as in the late election, not being able to correct existing evils among themselves, they determined by casting their ballots against those unprincipled adventurers, to overthrow them.... My people have been told by these schemers, when men have been placed on the ticket who were notoriously corrupt and dishonest, that they must vote for them; that the salvation of the party depended upon it; that the man who scratched a ticket was not a Republican. This is only one of the many means these unprincipled demagogues have devised to perpetuate the intellectual bondage of my people. To defeat this policy, at the late election men, irrespective of race, color, or party affiliation, united, and voted together against men known to be incompetent and dishonest. I cannot recognize, nor do the mass of my people who read, recognize the majority of the officials who have been in power for the past two years as Republican.*

Upon his exit from politics, from 1876 to 1882, Revels again served as the President of Alcorn University. He and his wife then moved to Holly Springs, Mississippi where he returned to the AME ministries. On January 16, 1901, Mr. Revels died of a stroke while attending the AME church conference in Aberdeen, Mississippi. Hiram Revels is buried in Hill Crest Cemetery in Holly Springs, Mississippi. Hiram Revels is known today for strong religious convictions, honesty, integrity and equality for all men.

Senator Hiram Rhodes Revels. Official portrait by Matthew Brady. (LOC)

Another famous African American politician from the Civil War era is Senator Bruce. Blanche Kelso Bruce (1841– 1898) was born a slave on a plantation in Farmville, Virginia. His mother, Polly Bruce, was a domestic slave and his father, Pettis Perkinson, was the owner of the plantation. Blanche Bruce was treated well and his father saw to his education as well. He attended Oberlin College in Ohio, and he taught there for several years following his graduation.

After college, and during the Civil War, Mr. Bruce worked as a steamboat porter on the Mississippi River and he settled in Hannibal, Missouri where he founded a school for black children. In 1868, Bruce moved to Bolivar, Mississippi where he bought a large tract of land and started a plantation.

During the Reconstruction Era, Bruce served in various civic positions including Registrar of Voters and County Tax Assessor. He also served as the Tax Collector, Education Supervisor and Sheriff of neighboring Bolivar County. In 1870, he served as the Sergeant at Arms in the Mississippi State Senate.

In 1874, Mr. Bruce was elected as a U.S. Republican Senator from Mississippi. He served from 1875 to 1881. During this time, in 1878, Blanche Bruce was married to an Ohio socialite and race activist, Josephine Beall Willson. They had one child, Roscoe Conkling Bruce. The Bruce family was very popular in societal affairs and they traveled the nation stumping for other politicians and various political causes. They also enjoyed entertaining and they took extended trips to Europe as well. Mrs. Bruce was instrumental in the formation of the National Association of Colored Women (NACW).

While Hiram Revels was the first African American to serve in the U.S. Congress, Senator Blanche Bruce was the first African American to serve a full term in the U.S. Congress.

In 1879, Senator Bruce served as the President of the U.S. Senate and accordingly, he was the first African American, and former slave to serve in that capacity. In 1881, President James A. Garfield appointed Senator Bruce as the head of the U.S. Treasury, and he was the first African American and former slave to have his signature appear on U.S. currency.

From 1890 to 1893, Mr. Bruce was appointed as the Recorder of Deeds in the District of Columbia. From 1892 to 1895, he also served on the District of Columbia Board of Trustees of Public Schools. In 1897, President William McKinley appointed Blanche Bruce to lead the Treasury again, and he served in that capacity until his death in 1898.

Following her husband's death, Mrs. Bruce served as the Dean of Women in the Tuskegee Institute in Alabama. She remained in that position until 1902. She died on February 24, 1923 at her son's home in Kendall, West Virginia.

Senator Blanche Kelso Bruce Official portrait by Matthew Brady. (LOC)

Following the war, important legislative acts during the Reconstruction Era provided important changes to the U.S. Constitution: the 13th Amendment, the abolition of slavery; the 14th Amendment, which defines citizenship and equal protection for all races; and the 15th Amendment, which ensures the right to vote for all races. Upon the ratification of these crucial civil rights laws, African Americans were afforded life-altering change, improved equality, and opportunity. For African Americans, the Civil War brought freedom, which serves as the first step in a long march that continues today.

Memorial Day Parade- *New York City, May 30, 1912, featuring Civil War veterans of the U.S. Army Colored Troops, Grand Army of the Republic.* (LOC)

Chapter 10

Civil War Timeline

There's many a boy here today who looks on war as all glory but it is all hell... War is cruelty. There is no use trying to reform it. The crueler it is, the sooner it will be over. War is the remedy that our enemies have chosen, and I say let us give them all they want.

– Union General William Tecumseh Sherman

November 6, 1860- Election of the sixteenth president of the United States. Abraham Lincoln, Republican wins. He opposed the spread of slavery in U.S. territories.

December 20, 1860- South Carolina secedes from the Union.

February 8-9, 1861 - Confederate States of America are formed with the government administration established at Montgomery, Alabama.

February 18, 1861- Jefferson Davis appointed President of the Confederate States of America.

March 4, 1861- Abraham Lincoln inauguration.

April 12, 1861- The Civil War is initiated as Southern troops bombard Fort Sumter, South Carolina.

April 15, 1861- President Lincoln declares an insurrection has formed in the South- calls for militia to quell the rebellion and expands the regular Army.

July 21, 1861- The Battle of Bull Run (or First Manassas) just west of Washington D.C., Southern General Beauregard defeats Union troops led by General McDowell.

February 22, 1862- Jefferson Davis secures formal vote as the President of the Confederate States of America.

August 5, 1864- The Battle of Mobile Bay of was a large scale naval battle. The Union fleet won a major victory which resulted in a successful Union blockade of the port for the remainder of the war. Four African Americans were cited for bravery for which they won the Medal of Honor.

September 17, 1862- The Battle of Antietam (or Sharpsburg), Maryland results in a Union victory and halts General Robert E. Lee's Confederate invasion into the North.

January 1, 1863- President Lincoln's Emancipation Proclamation is issued declaring all slaves are free men.

March 3, 1863- The draft (Conscription) begins in the Union Army.

July 1-3- The Battle of Gettysburg, Pennsylvania. Another key Union victory and Lee's Confederate troops are again halted in a northern offensive.

July 4- Vicksburg, Mississippi falls to Union forces opening the Mississippi to Union domination. In the east, Lee begins a long and bitter retreat from Gettysburg.

July 10-11, 1863- The Confederates in Charleston, South Carolina's heavily defended harbor region are threatened by Union naval and land attacks. The 54th Massachusetts Colored Infantry is engaged along with white Union troops. The action represents the first combat involving African American troops.

July 13-16, 1863- In New York City, the "Draft Riots" result in a violent uprising of men protesting the war and the draft. Many African Americans are lynched.

July 18, 1863- Assault on Battery Wagner, South Carolina. The 54th Massachusetts Colored Infantry leads the charge and although the attack is repelled, African Americans win great respect for their courage under fire. Sergeant William Carney becomes the first African American to be decorated with the Medal of Honor for valor exhibited in battle.

November 19, 1863- President Lincoln delivers the Gettysburg Address.

June 15-18, 1864- Union assault on the Confederate defenses guarding Petersburg, Virginia results in a stalemate and siege warfare begins.

July 21, 1864- The Siege of Atlanta begins as Union General Sherman presses a northern invasion against key Confederate defenses around Atlanta.

July 30, 1864- The Battle of the Crater at Petersburg, Virginia results in a mighty explosion initiated by Union troops who tunneled underneath Confederate entrenchments. While the explosive charges created a huge crater in the Confederate defenses, Union troops were halted in battle, and the siege continued.

September 1, 1864- Atlanta, Georgia is taken after a lengthy siege. General Hood evacuates his troops and General Sherman's army secures a key Confederate stronghold in the heart of the South.

September 29–30, 1864- At the Battles of Chaffin's Farm and New Market Heights, which also includes fighting at key Confederate defenses (Forts Harrison, Gilmer, and Johnson), an entire division of African American colored troops fight valiantly and secure key terrain. With the threat of CSA entrenchments in front of Richmond and Petersburg, General Lee is forced to spread his diminishing men and resources across a long, thin line. Fourteen African American soldiers are cited for valor and decoration with the Medal of Honor.

November 8, 1864- Abraham Lincoln secures a victory and is reelected as the president of the United States.

November 16, 1864- The "March to the Sea" is begun as General Sherman's Union Army cuts a path of destruction across the southern countryside. His initial goal is to wage war on a broad front moving towards Savannah.

April 2, 1865- By March 25th, Confederate forces could no longer defend the lengthy line of defenses and General Lee abandons both cities by the beginning of April. The Confederates begin a ragged retreat toward Appomattox Court House, Virginia. Lee's plan is to link up with additional troops there led by General Joseph Johnston.

April 3, 1865- Union troops fill the massive void left by retreating Confederate troops and they secure the cities of Richmond and Petersburg, Virginia.

April 9, 1865- Battle of Appomattox Court House, Virginia results in the surrender of General Lee's entire southern army to the victorious Union forces led by General U.S. Grant.

April 14, 1865- While watching a play at Ford's theatre in Washington D.C., John Wilkes Booth assassinates President Lincoln.

April 26, 1865- General Joseph Johnston formally surrenders his Confederate forces just outside Durham, North Carolina.

May 4, 1865- General Richard Taylor surrenders his Confederate forces in the Alabama, Mississippi and East Louisiana.

May 10, 1865- Confederate President Jefferson Davis is captured at Irwinville, Georgia while fleeing southward from Richmond.

May 12, 1865- The final battle of the Civil War is fought at Palmito Ranch, Texas.

May 23-24, 1865- The Grand Review parade of the Union Army is held in Washington, D.C.

May 24, 1865- The Grand Review of General Sherman's Army in Washington, DC

June 2, 1865- The final lot of Confederate forces, General Simon Bolivar Buckner surrenders his Army of the Trans-Mississippi, and the Civil War is officially ended.

1865- 1877- The Reconstruction era of the former Confederate states. During this time, Union troops occupied the Southern states, securing order, enforcing new federal laws of equality, and assisting the people in rebuilding destroyed infrastructures and establishing new government administrations.

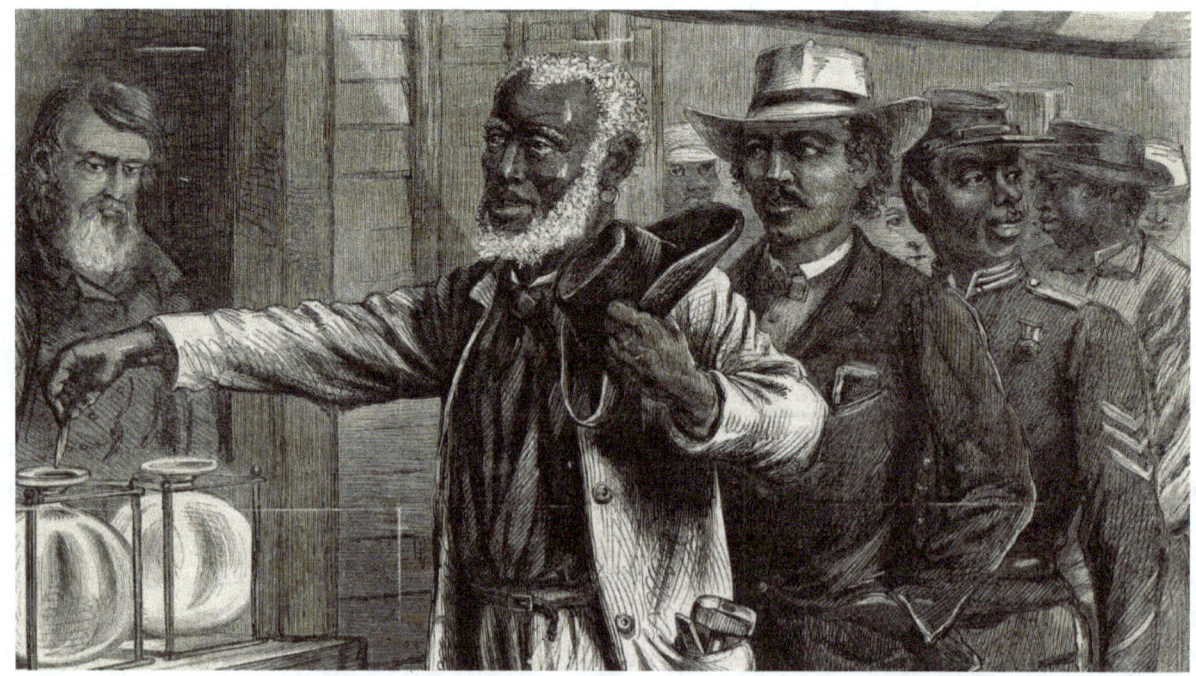

African Americans exercising their right to vote in elections. (Harper's Weekly)

Chapter 11

Selected Speeches and Publications

What to the Slave Is the Fourth of July?,"
by Frederick Douglass
(July 5, 1852)

Mr. President, Friends and Fellow Citizens:

He who could address this audience without a quailing sensation, has stronger nerves than I have. I do not remember ever to have appeared as a speaker before any assembly more shrinkingly, nor with greater distrust of my ability, than I do this day. A feeling has crept over me, quite unfavorable to the exercise of my limited powers of speech. The task before me is one which requires much previous thought and study for its proper performance.

I know that apologies of this sort are generally considered flat and unmeaning. I trust, however, that mine will not be so considered. Should I seem at ease, my appearance would much misrepresent me. The little experience I have had in addressing public meetings, in country schoolhouses, avails me nothing on the present occasion.

The papers and placards say, that I am to deliver a 4th [of] July oration. This certainly sounds large, and out of the common way, for it is true that I have often had the privilege to speak in this beautiful Hall, and to address many who now honor me with their presence. But neither their familiar faces, nor the perfect gage I think I have of Corinthian Hall, seems to free me from embarrassment.

The fact is, ladies and gentlemen, the distance between this platform and the slave plantation, from which I escaped, is considerable — and the difficulties to be overcome in getting from the latter to the former, are by no means slight. That I am here to-day is, to me, a matter of astonishment as well as of gratitude. You will not, therefore, be surprised, if in what I have to say I evince no elaborate preparation, nor grace my speech with any high-sounding exordium. With little experience and with less learning, I have been able to throw my thoughts hastily and imperfectly together; and trusting to your patient and generous indulgence, I will proceed to lay them before you.

This, for the purpose of this celebration, is the 4th of July. It is the birthday of your National Independence, and of your political freedom. This, to you, is what the Passover was to the emancipated people of God. It carries your minds back to the day, and to the act of your great deliverance; and to the signs, and to the wonders, associated with that act, and that day. This celebration also marks the beginning of another year of your national life; and reminds you that the Republic of America is now 76 years old. I am glad, fellow-citizens, that your nation is so young. Seventy-six years, though a good old age for a man, is but a mere speck in the life of a nation. Three score years and ten is the allotted time for individual men; but nations number their years by thousands. According to this fact, you are, even now, only in the beginning of your national career, still lingering in the period of childhood. I repeat, I am glad this is so.

There is hope in the thought, and hope is much needed, under the dark clouds which lower above the horizon. The eye of the reformer is met with angry flashes, portending disastrous times; but his heart may well beat lighter at the thought that America is young, and that she is still in the impressible stage of her existence. May he not hope that high lessons of wisdom, of justice and of truth, will yet give direction to her destiny? Were the nation older, the patriot's heart might be sadder, and the reformer's brow heavier. Its future might be shrouded in gloom, and the hope of its prophets go out in sorrow.

There is consolation in the thought that America is young. Great streams are not easily turned from channels, worn deep in the course of ages. They may sometimes rise in quiet and stately majesty, and inundate the land, refreshing and fertilizing the earth with their mysterious properties. They may also rise in wrath and fury, and bear away, on their angry waves, the accumulated wealth of years of toil and hardship. They, however, gradually flow back to the same old channel, and flow on as serenely as ever. But, while the river may not be turned aside, it may dry up, and leave nothing behind but the withered branch, and the unsightly rock, to howl in the abyss-sweeping wind, the sad tale of departed glory. As with rivers so with nations.

Fellow-citizens, I shall not presume to dwell at length on the associations that cluster about this day. The simple story of it is that, 76 years ago, the people of this country were British subjects. The style and title of your "sovereign people" (in which you now glory) was not then born. You were under the British Crown. Your fathers esteemed the English Government as the home government; and England as the fatherland. This home government, you know, although a considerable distance from your home, did, in the exercise of its parental prerogatives, impose upon its colonial children, such restraints, burdens and limitations, as, in its mature judgment, it deemed wise, right and proper.

But, your fathers, who had not adopted the fashionable idea of this day, of the infallibility of government, and the absolute character of its acts, presumed to differ from the home government in respect to the wisdom and the justice of some of those burdens and restraints. They went so far in their excitement as to pronounce the measures of government unjust, unreasonable, and oppressive, and altogether such as ought not to be quietly submitted to.

I scarcely need say, fellow-citizens, that my opinion of those measures fully accords with that of your fathers. Such a declaration of agreement on my part would not be worth much to anybody. It would, certainly, prove nothing, as to what part I might have taken, had I lived during the great controversy of 1776. To say now that America was right, and England wrong, is exceedingly easy. Everybody can say it; the dastard, not less than the noble brave, can flippantly discant on the tyranny of England towards the American Colonies.

It is fashionable to do so; but there was a time when to pronounce against England, and in favor of the cause of the colonies, tried men's souls. They who did so were accounted in their day, plotters of mischief, agitators and rebels, dangerous men. To side with the right, against the wrong, with the weak against the strong, and with the oppressed against the oppressor! here lies the merit, and the one which, of all others, seems unfashionable in our day. The cause of liberty may be stabbed by the men who glory in the deeds of your fathers. But, to proceed.

Feeling themselves harshly and unjustly treated by the home government, your fathers, like men of honesty, and men of spirit, earnestly sought redress. They petitioned and remonstrated; they did so in a decorous, respectful, and loyal manner. Their conduct was wholly unexceptionable. This, however, did not answer the purpose. They saw themselves treated with sovereign indifference, coldness and scorn. Yet they persevered. They were not the men to look back.

As the sheet anchor takes a firmer hold, when the ship is tossed by the storm, so did the cause of your fathers grow stronger, as it breasted the chilling blasts of kingly displeasure. The greatest and best of British statesmen admitted its justice, and the loftiest eloquence of the British Senate came to its support. But, with that blindness which seems to be the unvarying characteristic of tyrants, since Pharaoh and his hosts were drowned in the Red Sea, the British Government persisted in the exactions complained of.

The madness of this course, we believe, is admitted now, even by England; but we fear the lesson is wholly lost on our present ruler.

Oppression makes a wise man mad. Your fathers were wise men, and if they did not go mad, they became restive under this treatment. They felt themselves the victims of grievous wrongs, wholly incurable in their colonial capacity. With brave men there is always a remedy for oppression. Just here, the idea of a total separation of the colonies from the crown was born! It was a startling idea, much more so, than we, at this distance of time, regard it. The timid and the prudent (as has been intimated) of that day, were, of course, shocked and alarmed by it.

Such people lived then, had lived before, and will, probably, ever have a place on this planet; and their course, in respect to any great change, (no matter how great the good to be attained, or the wrong to be redressed by it), may be calculated with as much precision as can be the course of the stars. They hate all changes, but silver, gold and copper change! Of this sort of change they are always strongly in favor.

These people were called Tories in the days of your fathers; and the appellation, probably, conveyed the same idea that is meant by a more modern, though a somewhat less euphonious term, which we often find in our papers, applied to some of our old politicians.

Their opposition to the then dangerous thought was earnest and powerful; but, amid all their terror and affrighted vociferations against it, the alarming and revolutionary idea moved on, and the country with it.

On the 2d of July, 1776, the old Continental Congress, to the dismay of the lovers of ease, and the worshipers of property, clothed that dreadful idea with all the authority of national sanction. They did so in the form of a resolution; and as we seldom hit upon resolutions, drawn up in our day whose transparency is at all equal to this, it may refresh your minds and help my story if I read it. "Resolved, That these united colonies are, and of right, ought to be free and Independent States; that they are absolved from all allegiance to the British Crown; and that all political connection between them and the State of Great Britain is, and ought to be, dissolved."

Citizens, your fathers made good that resolution. They succeeded; and to-day you reap the fruits of their success. The freedom gained is yours; and you, therefore, may properly celebrate this anniversary. The 4th of July is the first great fact in your nation's history — the very ring-bolt in the chain of your yet undeveloped destiny.

Pride and patriotism, not less than gratitude, prompt you to celebrate and to hold it in perpetual remembrance. I have said that the Declaration of Independence is the ring-bolt to the chain of your nation's destiny; so, indeed, I regard it. The principles contained in that instrument are saving principles. Stand by those principles, be true to them on all occasions, in all places, against all foes, and at whatever cost.

From the round top of your ship of state, dark and threatening clouds may be seen. Heavy billows, like mountains in the distance, disclose to the leeward huge forms of flinty rocks! That bolt drawn, that chain broken, and all is lost. Cling to this day — cling to it, and to its principles, with the grasp of a storm-tossed mariner to a spar at midnight.

The coming into being of a nation, in any circumstances, is an interesting event. But, besides general considerations, there were peculiar circumstances which make the advent of this republic an event of special attractiveness.

The whole scene, as I look back to it, was simple, dignified and sublime.

The population of the country, at the time, stood at the insignificant number of three millions. The country was poor in the munitions of war. The population was weak and scattered, and the country a wilderness unsubdued. There were then no means of concert and combination, such as exist now. Neither steam nor lightning had then been reduced to order and discipline. From the Potomac to the Delaware was a journey of many days. Under these, and innumerable other disadvantages, your fathers declared for liberty and independence and triumphed.

Fellow Citizens, I am not wanting in respect for the fathers of this republic. The signers of the Declaration of Independence were brave men. They were great men too — great enough to give fame to a great age. It does not often happen to a nation to raise, at one time, such a number of truly great men. The point from which I am compelled to view them is not, certainly, the most favorable; and yet I cannot contemplate their great deeds with less than admiration. They were statesmen, patriots and heroes, and for the good they did, and the principles they contended for, I will unite with you to honor their memory.

They loved their country better than their own private interests; and, though this is not the highest form of human excellence, all will concede that it is a rare virtue, and that when it is exhibited, it ought to command respect. He who will, intelligently, lay down his life for his country, is a man whom it is not in human nature to despise. Your fathers staked their lives, their fortunes, and their sacred honor, on the cause of their country. In their admiration of liberty, they lost sight of all other interests.

They were peace men; but they preferred revolution to peaceful submission to bondage. They were quiet men; but they did not shrink from agitating against oppression. They showed forbearance; but that they knew its limits. They believed in order; but not in the order of tyranny. With them, nothing was "settled" that was not right. With them, justice, liberty and humanity were "final;" not slavery and oppression. You may well cherish the memory of such men. They were great in their day and generation. Their solid manhood stands out the more as we contrast it with these degenerate times.

How circumspect, exact and proportionate were all their movements! How unlike the politicians of an hour! Their statesmanship looked beyond the passing moment, and stretched away in strength into the distant future. They seized upon eternal principles, and set a glorious example in their defense. Mark them!

Fully appreciating the hardship to be encountered, firmly believing in the right of their cause, honorably inviting the scrutiny of an on-looking world, reverently appealing to heaven to attest their sincerity, soundly comprehending the solemn responsibility they were about to assume, wisely measuring the terrible odds against them, your fathers, the fathers of this republic, did, most deliberately, under the inspiration of a glorious patriotism, and with a sublime faith in the great principles of justice and freedom, lay deep the corner-stone of the national superstructure, which has risen and still rises in grandeur around you.

Of this fundamental work, this day is the anniversary. Our eyes are met with demonstrations of joyous enthusiasm. Banners and pennants wave exultingly on the breeze. The din of business, too, is hushed. Even Mammon seems to have quitted his grasp on this day. The ear-piercing fife and the stirring drum unite their accents with the ascending peal of a thousand church bells. Prayers are made, hymns are sung, and sermons are preached in honor of this day; while the quick martial tramp of a great and multitudinous nation, echoed back by all the hills, valleys and mountains of a vast continent, bespeak the occasion one of thrilling and universal interest — a nation's jubilee.

Friends and citizens, I need not enter further into the causes which led to this anniversary. Many of you understand them better than I do. You could instruct me in regard to them. That is a branch of knowledge in which you feel, perhaps, a much deeper interest than your speaker. The causes which led to the separation of the colonies from the British crown have never lacked for a tongue. They have all been taught in your common schools, narrated at your firesides, unfolded from your pulpits, and thundered from your legislative halls, and are as familiar to you as household words. They form the staple of your national poetry and eloquence.

I remember, also, that, as a people, Americans are remarkably familiar with all facts which make in their own favor. This is esteemed by some as a national trait — perhaps a national weakness. It is a fact, that whatever makes for the wealth or for the reputation of Americans, and can be had cheap! will be found by Americans. I shall not be charged with slandering Americans, if I say I think the American side of any question may be safely left in American hands.

I leave, therefore, the great deeds of your fathers to other gentlemen whose claim to have been regularly descended will be less likely to be disputed than mine!

My business, if I have any here to-day, is with the present. The accepted time with God and his cause is the ever-living now.

Trust no future, however pleasant,
Let the dead past bury its dead;
Act, act in the living present,
Heart within, and God overhead.

We have to do with the past only as we can make it useful to the present and to the future. To all inspiring motives, to noble deeds which can be gained from the past, we are welcome. But now is the time, the important time. Your fathers have lived, died, and have done their work, and have done much of it well. You live and must die, and you must do your work.

You have no right to enjoy a child's share in the labor of your fathers, unless your children are to be blest by your labors. You have no right to wear out and waste the hard-earned fame of your fathers to cover your indolence. Sydney Smith tells us that men seldom eulogize the wisdom and virtues of their fathers, but to excuse some folly or wickedness of their own. This truth is not a doubtful one. There are illustrations of it near and remote, ancient and modern.

It was fashionable, hundreds of years ago, for the children of Jacob to boast, we have "Abraham to our father," when they had long lost Abraham's faith and spirit. That people contented themselves under the shadow of Abraham's great name, while they repudiated the deeds which made his name great. Need I remind you that a similar thing is being done all over this country to-day? Need I tell you that the Jews are not the only people who built the tombs of the prophets, and garnished the sepulchres of the righteous? Washington could not die till he had broken the chains of his slaves. Yet his monument is built up by the price of human blood, and the traders in the bodies and souls of men shout — "We have Washington to *our father*." — Alas! that it should be so; yet so it is.

The evil that men do, lives after them, The good is oft-interred with their bones.

Fellow-citizens, pardon me, allow me to ask, why am I called upon to speak here to-day? What have I, or those I represent, to do with your national independence? Are the great principles of political freedom and of natural justice, embodied in that Declaration of Independence, extended to us? and am I, therefore, called upon to bring our humble offering to the national altar, and to confess the benefits and express devout gratitude for the blessings resulting from your independence to us?

Would to God, both for your sakes and ours, that an affirmative answer could be truthfully returned to these questions! Then would my task be light, and my burden easy and delightful. For who is there so cold, that a nation's sympathy could not warm him? Who so obdurate and dead to the claims of gratitude, that would not thankfully acknowledge such priceless benefits? Who so stolid and selfish, that would not give his voice to swell the hallelujahs of a nation's jubilee, when the chains of servitude had been torn from his limbs? I am not that man. In a case like that, the dumb might eloquently speak, and the "lame man leap as an hart."

But, such is not the state of the case. I say it with a sad sense of the disparity between us. I am not included within the pale of this glorious anniversary! Your high independence only reveals the immeasurable distance between us. The blessings in which you, this day, rejoice, are not enjoyed in common. — The rich inheritance of justice, liberty, prosperity and independence, bequeathed by your fathers, is shared by you, not by me. The sunlight that brought life and healing to you, has brought stripes and death to me. This Fourth [of] July is *yours*, not *mine*.

You may rejoice, *I* must mourn. To drag a man in fetters into the grand illuminated temple of liberty, and call upon him to join you in joyous anthems, were inhuman mockery and sacrilegious irony. Do you mean, citizens, to mock me, by asking me to speak to-day? If so, there is a parallel to your conduct. And let me warn you that it is dangerous to copy the example of a nation whose crimes, lowering up to heaven, were thrown down by the breath of the Almighty, burying that nation in irrecoverable ruin! I can to-day take up the plaintive lament of a peeled and woe-smitten people!

"By the rivers of Babylon, there we sat down. Yea! we wept when we remembered Zion. We hanged our harps upon the willows in the midst thereof. For there, they that carried us away captive, required of us a song; and they who wasted us required of us mirth, saying, Sing us one of the songs of Zion. How can we sing the Lord's song in a strange land? If I forget thee, O Jerusalem, let my right hand forget her cunning. If I do not remember thee, let my tongue cleave to the roof of my mouth."

Fellow-citizens; above your national, tumultuous joy, I hear the mournful wail of millions! whose chains, heavy and grievous yesterday, are, to-day, rendered more intolerable by the jubilee shouts that reach them. If I do forget, if I do not faithfully remember those bleeding children of sorrow this day, "may my right hand forget her cunning, and may my tongue cleave to the roof of my mouth!" To forget them, to pass lightly over their wrongs, and to chime in with the popular theme, would be treason most scandalous and shocking, and would make me a reproach before God and the world. My subject, then fellow-citizens, is AMERICAN SLAVERY.

I shall see, this day, and its popular characteristics, from the slave's point of view. Standing, there, identified with the American bondman, making his wrongs mine, I do not hesitate to declare, with all my soul, that the character and conduct of this nation never looked blacker to me than on this 4th of July! Whether we turn to the declarations of the past, or to the professions of the present, the conduct of the nation seems equally hideous and revolting. America is false to the past, false to the present, and solemnly binds herself to be false to the future.

Standing with God and the crushed and bleeding slave on this occasion, I will, in the name of humanity which is outraged, in the name of liberty which is fettered, in the name of the constitution and the Bible, which are disregarded and trampled upon, dare to call in question and to denounce, with all the emphasis I can command, everything that serves to perpetuate slavery — the great sin and shame of America! "I will not equivocate; I will not excuse;" I will use the severest language I can command; and yet not one word shall escape me that any man, whose judgment is not blinded by prejudice, or who is not at heart a slaveholder, shall not confess to be right and just.

But I fancy I hear some one of my audience say, it is just in this circumstance that you and your brother abolitionists fail to make a favorable impression on the public mind. Would you argue more, and denounce less, would you persuade more, and rebuke less, your cause would be much more likely to succeed. But, I submit, where all is plain there is nothing to be argued. What point in the anti-slavery creed would you have me argue? On what branch of the subject do the people of this country need light? Must I undertake to prove that the slave is a man? That point is conceded already. Nobody doubts it.

The slaveholders themselves acknowledge it in the enactment of laws for their government. They acknowledge it when they punish disobedience on the part of the slave. There are seventy-two crimes in the State of Virginia, which, if committed by a black man, (no matter how ignorant he be), subject him to the punishment of death; while only two of the same crimes will subject a white man to the like punishment. What is this but the acknowledgement that the slave is a moral, intellectual and responsible being? The manhood of the slave is conceded. It is admitted in the fact that Southern statute books are covered with enactments forbidding, under severe fines and penalties, the teaching of the slave to read or to write.

When you can point to any such laws, in reference to the beasts of the field, then I may consent to argue the manhood of the slave. When the dogs in your streets, when the fowls of the air, when the cattle on your hills, when the fish of the sea, and the reptiles that crawl, shall be unable to distinguish the slave from a brute, *then* will I argue with you that the slave is a man!

For the present, it is enough to affirm the equal manhood of the Negro race. Is it not astonishing that, while we are ploughing, planting and reaping, using all kinds of mechanical tools, erecting houses, constructing bridges, building ships, working in metals of brass, iron, copper, silver and gold; that, while we are reading, writing and cyphering, acting as clerks, merchants and secretaries, having among us lawyers, doctors, ministers, poets, authors, editors, orators and teachers; that, while we are engaged in all manner of enterprises common to other men, digging gold in California, capturing the whale in the Pacific, feeding sheep and cattle on the hill-side, living, moving, acting, thinking, planning, living in families as husbands, wives and children, and, above all, confessing and worshipping the Christian's God, and looking hopefully for life and immortality beyond the grave, we are called upon to prove that we are men!

Would you have me argue that man is entitled to liberty? that he is the rightful owner of his own body? You have already declared it. Must I argue the wrongfulness of slavery? Is that a question for Republicans? Is it to be settled by the rules of logic and argumentation, as a matter beset with great difficulty, involving a doubtful application of the principle of justice, hard to be understood? How should I look to-day, in the presence of Americans, dividing, and subdividing a discourse, to show that men have a natural right to freedom? speaking of it relatively, and positively, negatively, and affirmatively. To do so, would be to make myself ridiculous, and to offer an insult to your understanding. — There is not a man beneath the canopy of heaven, that does not know that slavery is wrong *for him*.

What, am I to argue that it is wrong to make men brutes, to rob them of their liberty, to work them without wages, to keep them ignorant of their relations to their fellow men, to beat them with sticks, to flay their flesh with the lash, to load their limbs with irons, to hunt them with dogs, to sell them at auction, to sunder their families, to knock out their teeth, to burn their flesh, to starve them into obedience and submission to their masters? Must I argue that a system thus marked with blood, and stained with pollution, is *wrong*? No! I will not. I have better employments for my time and strength than such arguments would imply.

What, then, remains to be argued? Is it that slavery is not divine; that God did not establish it; that our doctors of divinity are mistaken? There is blasphemy in the thought. That which is inhuman, cannot be divine! Who can reason on such a proposition? They that can, may; I cannot. The time for such argument is passed.

At a time like this, scorching irony, not convincing argument, is needed. O! had I the ability, and could I reach the nation's ear, I would, to-day, pour out a fiery stream of biting ridicule, blasting reproach, withering sarcasm, and stern rebuke. For it is not light that is needed, but fire; it is not the gentle shower, but thunder. We need the storm, the whirlwind, and the earthquake. The feeling of the nation must be quickened; the conscience of the nation must be roused; the propriety of the nation must be startled; the hypocrisy of the nation must be exposed; and its crimes against God and man must be proclaimed and denounced.

What, to the American slave, is your 4th of July? I answer: a day that reveals to him, more than all other days in the year, the gross injustice and cruelty to which he is the constant victim. To him, your celebration is a sham; your boasted liberty, an unholy license; your national greatness, swelling vanity; your sounds of rejoicing are empty and heartless; your denunciations of tyrants, brass fronted impudence; your shouts of liberty and equality, hollow mockery; your prayers and hymns, your sermons and thanksgivings, with all your religious parade, and solemnity, are, to him, mere bombast, fraud, deception, impiety, and hypocrisy — a thin veil to cover up crimes which would disgrace a nation of savages. There is not a nation on the earth guilty of practices, more shocking and bloody, than are the people of these United States, at this very hour.

Go where you may, search where you will, roam through all the monarchies and despotisms of the old world, travel through South America, search out every abuse, and when you have found the last, lay your facts by the side of the everyday practices of this nation, and you will say with me, that, for revolting barbarity and shameless hypocrisy, America reigns without a rival.

Take the American slave-trade, which, we are told by the papers, is especially prosperous just now. Ex-Senator Benton tells us that the price of men was never higher than now. He mentions the fact to show that slavery is in no danger. This trade is one of the peculiarities of American institutions. It is carried on in all the large towns and cities in one-half of this confederacy; and millions are pocketed every year, by dealers in this horrid traffic. In several states, this trade is a chief source of wealth. It is called (in contradistinction to the foreign slave-trade) "*the internal slave trade.*" It is, probably, called so, too, in order to divert from it the horror with which the foreign slave-trade is contemplated. That trade has long since been denounced by this government, as piracy.

It has been denounced with burning words, from the high places of the nation, as an execrable traffic. To arrest it, to put an end to it, this nation keeps a squadron, at immense cost, on the coast of Africa. Everywhere, in this country, it is safe to speak of this foreign slave-trade, as a most inhuman traffic, opposed alike to the laws of God and of man. The duty to extirpate and destroy it, is admitted even by our DOCTORS OF DIVINITY.

In order to put an end to it, some of these last have consented that their colored brethren (nominally free) should leave this country, and establish themselves on the western coast of Africa! It is, however, a notable fact that, while so much execration is poured out by Americans upon those engaged in the foreign slave-trade, the men engaged in the slave-trade between the states pass without condemnation, and their business is deemed honorable.

Behold the practical operation of this internal slave-trade, the American slave-trade, sustained by American politics and America religion. Here you will see men and women reared like swine for the market. You know what is a swine-drover? I will show you a man-drover. They inhabit all our Southern States. They perambulate the country, and crowd the highways of the nation, with droves of human stock. You will see one of these human flesh-jobbers, armed with pistol, whip and bowie-knife, driving a company of a hundred men, women, and children, from the Potomac to the slave market at New Orleans.

These wretched people are to be sold singly, or in lots, to suit purchasers. They are food for the cotton-field, and the deadly sugar-mill. Mark the sad procession, as it moves wearily along, and the inhuman wretch who drives them. Hear his savage yells and his blood-chilling oaths, as he hurries on his affrighted captives! There, see the old man, with locks thinned and gray. Cast one glance, if you please, upon that young mother, whose shoulders are bare to the scorching sun, her briny tears falling on the brow of the babe in her arms.

See, too, that girl of thirteen, weeping, *yes*! weeping, as she thinks of the mother from whom she has been torn! The drove moves tardily. Heat and sorrow have nearly consumed their strength; suddenly you hear a quick snap, like the discharge of a rifle; the fetters clank, and the chain rattles simultaneously; your ears are saluted with a scream, that seems to have torn its way to the center of your soul! The crack you heard, was the sound of the slave-whip; the scream you heard, was from the woman you saw with the babe. Her speed had faltered under the weight of her child and her chains! that gash on her shoulder tells her to move on.

Follow the drove to New Orleans. Attend the auction; see men examined like horses; see the forms of women rudely and brutally exposed to the shocking gaze of American slave-buyers. See this drove sold and separated forever; and never forget the deep, sad sobs that arose from that scattered multitude. Tell me citizens, WHERE, under the sun, you can witness a spectacle more fiendish and shocking. Yet this is but a glance at the American slave-trade, as it exists, at this moment, in the ruling part of the United States.

I was born amid such sights and scenes. To me the American slave-trade is a terrible reality. When a child, my soul was often pierced with a sense of its horrors. I lived on Philpot Street, Fell's Point, Baltimore, and have watched from the wharves, the slave ships in the Basin, anchored from the shore, with their cargoes of human flesh, waiting for favorable winds to waft them down the Chesapeake. There was, at that time, a grand slave mart kept at the head of Pratt Street, by Austin Woldfolk. His agents were sent into every town and county in Maryland, announcing their arrival, through the papers, and on flaming "*hand-bills*," headed CASH FOR NEGROES.

These men were generally well dressed men, and very captivating in their manners. Ever ready to drink, to treat, and to gamble. The fate of many a slave has depended upon the turn of a single card; and many a child has been snatched from the arms of its mother by bargains arranged in a state of brutal drunkenness.

The flesh-mongers gather up their victims by dozens, and drive them, chained, to the general depot at Baltimore. When a sufficient number have been collected here, a ship is chartered, for the purpose of conveying the forlorn crew to Mobile, or to New Orleans. From the slave prison to the ship, they are usually driven in the darkness of night; for since the antislavery agitation, a certain caution is observed.

In the deep still darkness of midnight, I have been often aroused by the dead heavy footsteps, and the piteous cries of the chained gangs that passed our door. The anguish of my boyish heart was intense; and I was often consoled, when speaking to my mistress in the morning, to hear her say that the custom was very wicked; that she hated to hear the rattle of the chains, and the heart-rending cries. I was glad to find one who sympathized with me in my horror.

Fellow-citizens, this murderous traffic is, to-day, in active operation in this boasted republic. In the solitude of my spirit, I see clouds of dust raised on the highways of the South; I see the bleeding footsteps; I hear the doleful wail of fettered humanity, on the way to the slave-markets, where the victims are to be sold like *horses*, *sheep*, and *swine*, knocked off to the highest bidder. There I see the tenderest ties ruthlessly broken, to gratify the lust, caprice and rapacity of the buyers and sellers of men. My soul sickens at the sight.

Is this the land your Fathers loved,
The freedom which they toiled to win?
Is this the earth whereon they moved?
Are these the graves they slumber in?

But a still more inhuman, disgraceful, and scandalous state of things remains to be presented. By an act of the American Congress, not yet two years old, slavery has been nationalized in its most horrible and revolting form. By that act, Mason and Dixon's line has been obliterated; New York has become as Virginia; and the power to hold, hunt, and sell men, women, and children as slaves remains no longer a mere state institution, but is now an institution of the whole United States.

The power is co-extensive with the Star-Spangled Banner and American Christianity. Where these go, may also go the merciless slave-hunter. Where these are, man is not sacred. He is a bird for the sportsman's gun. By that most foul and fiendish of all human decrees, the liberty and person of every man are put in peril. Your broad republican domain is hunting ground for *men*. Not for thieves and robbers, enemies of society, merely, but for men guilty of no crime. Your lawmakers have commanded all good citizens to engage in this hellish sport.

Your President, your Secretary of State, our *lords*, *nobles*, and ecclesiastics, enforce, as a duty you owe to your free and glorious country, and to your God, that you do this accursed thing. Not fewer than forty Americans have, within the past two years, been hunted down and, without a moment's warning, hurried away in chains, and consigned to slavery and excruciating torture. Some of these have had wives and children, dependent on them for bread; but of this, no account was made. The right of the hunter to his prey stands superior to the right of marriage, and to *all* rights in this republic, the rights of God included!

For black men there are neither law, justice, humanity, not religion. The Fugitive Slave *Law* makes mercy to them a crime; and bribes the judge who tries them. An American judge gets ten dollars for every victim he consigns to slavery, and five, when he fails to do so. The oath of any two villains is sufficient, under this hell-black enactment, to send the most pious and exemplary black man into the remorseless jaws of slavery! His own testimony is nothing. He can bring no witnesses for himself.

The minister of American justice is bound by the law to hear but *one* side; and *that* side, is the side of the oppressor. Let this damning fact be perpetually told. Let it be thundered around the world, that, in tyrant-killing, king-hating, people-loving, democratic, Christian America, the seats of justice are filled with judges, who hold their offices under an open and palpable *bribe*, and are bound, in deciding in the case of a man's liberty, *hear only his accusers*!

In glaring violation of justice, in shameless disregard of the forms of administering law, in cunning arrangement to entrap the defenseless, and in diabolical intent, this Fugitive Slave Law stands alone in the annals of tyrannical legislation. I doubt if there be another nation on the globe, having the brass and the baseness to put such a law on the statute-book. If any man in this assembly thinks differently from me in this matter, and feels able to disprove my statements, I will gladly confront him at any suitable time and place he may select.

I take this law to be one of the grossest infringements of Christian Liberty, and, if the churches and ministers of our country were not stupidly blind, or most wickedly indifferent, they, too, would so regard it.

At the very moment that they are thanking God for the enjoyment of civil and religious liberty, and for the right to worship God according to the dictates of their own consciences, they are utterly silent in respect to a law which robs religion of its chief significance, and makes it utterly worthless to a world lying in wickedness. Did this law concern the "*mint, anise, and cumin*" — abridge the right to sing psalms, to partake of the sacrament, or to engage in any of the ceremonies of religion, it would be smitten by the thunder of a thousand pulpits. A general shout would go up from the church, demanding *repeal, repeal, instant repeal*! — And it would go hard with that politician who presumed to solicit the votes of the people without inscribing this motto on his banner. Further, if this demand were not complied with, another Scotland would be added to the history of religious liberty, and the stern old Covenanters would be thrown into the shade. A John Knox would be seen at every church door, and heard from every pulpit, and Fillmore would have no more quarter than was shown by Knox, to the beautiful, but treacherous queen Mary of Scotland.

The fact that the church of our country, (with fractional exceptions), does not esteem "the Fugitive Slave Law" as a declaration of war against religious liberty, implies that that church regards religion simply as a form of worship, an empty ceremony, and *not* a vital principle, requiring active benevolence, justice, love and good will towards man. It esteems sacrifice above mercy; psalm-singing above right doing; solemn meetings above practical righteousness.

A worship that can be conducted by persons who refuse to give shelter to the houseless, to give bread to the hungry, clothing to the naked, and who enjoin obedience to a law forbidding these acts of mercy, is a curse, not a blessing to mankind. The Bible addresses all such persons as "scribes, Pharisees, hypocrites, who pay tithe of *mint, anise*, and *cumin*, and have omitted the weightier matters of the law, judgment, mercy and faith."

But the church of this country is not only indifferent to the wrongs of the slave, it actually takes sides with the oppressors. It has made itself the bulwark of American slavery, and the shield of American slave-hunters. Many of its most eloquent Divines. who stand as the very lights of the church, have shamelessly given the sanction of religion and the Bible to the whole slave system. They have taught that man may, properly, be a slave; that the relation of master and slave is ordained of God; that to send back an escaped bondman to his master is clearly the duty of all the followers of the Lord Jesus Christ; and this horrible blasphemy is palmed off upon the world for Christianity.

For my part, I would say, welcome infidelity! welcome atheism! welcome anything! in preference to the gospel, *as preached by those Divines*! They convert the very name of religion into an engine of tyranny, and barbarous cruelty, and serve to confirm more infidels, in this age, than all the infidel writings of Thomas Paine, Voltaire, and Bolingbroke, put together, have done! These ministers make religion a cold and flinty-hearted thing, having neither principles of right action, nor bowels of compassion. They strip the love of God of its beauty, and leave the throng of religion a huge, horrible, repulsive form. It is a religion for oppressors, tyrants, man-stealers, and *thugs*.

It is not that "*pure and undefiled religion*" which is from above, and which is "*first pure, then peaceable, easy to be entreated,* full of mercy and good fruits, *without partiality, and without hypocrisy*." But a religion which favors the rich against the poor; which exalts the proud above the humble; which divides mankind into two classes, tyrants and slaves; which says to the man in chains, *stay there*; and to the oppressor, *oppress on*; it is a religion which may be professed and enjoyed by all the robbers and enslavers of mankind; it makes God a respecter of persons, denies his fatherhood of the race, and tramples in the dust the great truth of the brotherhood of man. All this we affirm to be true of the popular church, and the popular worship of our land and nation — a religion, a church, and a worship which, on the authority of inspired wisdom, we pronounce to be an abomination in the sight of God.

In the language of Isaiah, the American church might be well addressed, "Bring no more vain ablations; incense is an abomination unto me: the new moons and Sabbaths, the calling of assemblies,

I cannot away with; it is iniquity even the solemn meeting. Your new moons and your appointed feasts my soul hateth. They are a trouble to me; I am weary to bear them; and when ye spread forth your hands I will hide mine eyes from you. Yea! when ye make many prayers, I will not hear. YOUR HANDS ARE FULL OF BLOOD; cease to do evil, learn to do well; seek judgment; relieve the oppressed; judge for the fatherless; plead for the widow."

The American church is guilty, when viewed in connection with what it is doing to uphold slavery; but it is superlatively guilty when viewed in connection with its ability to abolish slavery. The sin of which it is guilty is one of omission as well as of commission. Albert Barnes but uttered what the common sense of every man at all observant of the actual state of the case will receive as truth, when he declared that "There is no power out of the church that could sustain slavery an hour, if it were not sustained in it."

Let the religious press, the pulpit, the Sunday school, the conference meeting, the great ecclesiastical, missionary, Bible and tract associations of the land array their immense powers against slavery and slave-holding; and the whole system of crime and blood would be scattered to the winds; and that they do not do this involves them in the most awful responsibility of which the mind can conceive.

In prosecuting the anti-slavery enterprise, we have been asked to spare the church, to spare the ministry; but *how*, we ask, could such a thing be done? We are met on the threshold of our efforts for the redemption of the slave, by the church and ministry of the country, in battle arrayed against us; and we are compelled to fight or flee. From *what* quarter, I beg to know, has proceeded a fire so deadly upon our ranks, during the last two years, as from the Northern pulpit? As the champions of oppressors, the chosen men of American theology have appeared — men, honored for their so-called piety, and their real learning.

The Lords of Buffalo, the Springs of New York, the Lathrops of Auburn, the Coxes and Spencers of Brooklyn, the Gannets and Sharps of Boston, the Deweys of Washington, and other great religious lights of the land have, in utter denial of the authority of *Him* by whom they professed to be called to the ministry, deliberately taught us, against the example or the Hebrews and against the remonstrance of the Apostles, they teach *that we ought to obey man's law before the law of God.*

My spirit wearies of such blasphemy; and how such men can be supported, as the "standing types and representatives of Jesus Christ," is a mystery which I leave others to penetrate. In speaking of the American church, however, let it be distinctly understood that I mean the great mass of the religious organizations of our land. There are exceptions, and I thank God that there are. Noble men may be found, scattered all over these Northern States, of whom Henry Ward Beecher of Brooklyn, Samuel J. May of Syracuse, and my esteemed friend (Rev. R. R. Raymond) on the platform, are shining examples; and let me say further, that upon these men lies the duty to inspire our ranks with high religious faith and zeal, and to cheer us on in the great mission of the slave's redemption from his chains.

One is struck with the difference between the attitude of the American church towards the anti-slavery movement, and that occupied by the churches in England towards a similar movement in that country. There, the church, true to its mission of ameliorating, elevating, and improving the condition of mankind, came forward promptly, bound up the wounds of the West Indian slave, and restored him to his liberty. There, the question of emancipation was a high religious question. It was demanded, in the name of humanity, and according to the law of the living God.

The Sharps, the Clarksons, the Wilberforces, the Buxtons, and Burchells and the Knibbs, were alike famous for their piety, and for their philanthropy. The anti-slavery movement *there* was not an anti-church movement, for the reason that the church took its full share in prosecuting that movement: and the anti-slavery movement in this country will cease to be an anti-church movement, when the church of this country shall assume a favorable, instead of a hostile position towards that movement. Americans! your republican politics, not less than your republican religion, are flagrantly inconsistent.

You boast of your love of liberty, your superior civilization, and your pure Christianity, while the whole political power of the nation (as embodied in the two great political parties), is solemnly pledged to support and perpetuate the enslavement of three millions of your countrymen. You hurl your anathemas at the crowned headed tyrants of Russia and Austria, and pride yourselves on your Democratic institutions, while you yourselves consent to be the mere *tools* and *body-guards* of the tyrants of Virginia and Carolina.

You invite to your shores fugitives of oppression from abroad, honor them with banquets, greet them with ovations, cheer them, toast them, salute them, protect them, and pour out your money to them like water; but the fugitives from your own land you advertise, hunt, arrest, shoot and kill. You glory in your refinement and your universal education yet you maintain a system as barbarous and dreadful as ever stained the character of a nation — a system begun in avarice, supported in pride, and perpetuated in cruelty.

You shed tears over fallen Hungary, and make the sad story of her wrongs the theme of your poets, statesmen and orators, till your gallant sons are ready to fly to arms to vindicate her cause against her oppressors; but, in regard to the ten thousand wrongs of the American slave, you would enforce the strictest silence, and would hail him as an enemy of the nation who dares to make those wrongs the subject of public discourse! You are all on fire at the mention of liberty for France or for Ireland; but are as cold as an iceberg at the thought of liberty for the enslaved of America.

You discourse eloquently on the dignity of labor; yet, you sustain a system which, in its very essence, casts a stigma upon labor. You can bare your bosom to the storm of British artillery to throw off a three-penny tax on tea; and yet wring the last hard-earned farthing from the grasp of the black laborers of your country. You profess to believe "that, of one blood, God made all nations of men to dwell on the face of all the earth," and hath commanded all men, everywhere to love one another; yet you notoriously hate, (and glory in your hatred), all men whose skins are not colored like your own.

You declare, before the world, and are understood by the world to declare, that you *"hold these truths to be self-evident, that all men are created equal; and are endowed by their Creator with certain inalienable rights; and that, among these are, life, liberty, and the pursuit of happiness;"* and yet, you hold securely, in a bondage which, according to your own Thomas Jefferson, *"is worse than ages of that which your fathers rose in rebellion to oppose,"* a *seventh part* of the inhabitants of your country.

Fellow-citizens! I will not enlarge further on your national inconsistencies. The existence of slavery in this country brands your republicanism as a sham, your humanity as a base pretense, and your Christianity as a lie. It destroys your moral power abroad; it corrupts your politicians at home. It saps the foundation of religion; it makes your name a hissing, and a bye-word to a mocking earth. It is the antagonistic force in your government, the only thing that seriously disturbs and endangers your *Union*. It fetters your progress; it is the enemy of improvement, the deadly foe of education; it fosters pride; it breeds insolence; it promotes vice; it shelters crime; it is a curse to the earth that supports it; and yet, you cling to it, as if it were the sheet anchor of all your hopes. Oh! be warned! be warned! a horrible reptile is coiled up in your nation's bosom; the venomous creature is nursing at the tender breast of your youthful republic; *for the love of God*, tear away, and fling from you the hideous monster, and *let the weight of twenty millions crush and destroy it forever*!

But it is answered in reply to all this, that precisely what I have now denounced is, in fact, guaranteed and sanctioned by the Constitution of the United States; that the right to hold and to hunt slaves is a part of that Constitution framed by the illustrious Fathers of this Republic.

Then, I dare to affirm, notwithstanding all I have said before, your fathers stooped, basely stooped

To palter with us in a double sense:
And keep the word of promise to the ear,
But break it to the heart.

And instead of being the honest men I have before declared them to be, they were the veriest imposters that ever practiced on mankind. This is the inevitable conclusion, and from it there is no escape. But I differ from those who charge this baseness on the framers of the Constitution of the United States. It is a slander upon their memory, at least, so I believe. There is not time now to argue the constitutional question at length — nor have I the ability to discuss it as it ought to be discussed. The subject has been handled with masterly power by Lysander Spooner, Esq., by William Goodell, by Samuel E. Sewall, Esq., and last, though not least, by Gerritt Smith, Esq. These gentlemen have, as I think, fully and clearly vindicated the Constitution from any design to support slavery for an hour.

Fellow-citizens! there is no matter in respect to which, the people of the North have allowed themselves to be so ruinously imposed upon, as that of the pro-slavery character of the Constitution. In that instrument I hold there is neither warrant, license, nor sanction of the hateful thing; but, interpreted as it ought to be interpreted, the Constitution is a GLORIOUS LIBERTY DOCUMENT. Read its preamble, consider its purposes. Is slavery among them? Is it at the gateway? or is it in the temple? It is neither.

While I do not intend to argue this question on the present occasion, let me ask, if it be not somewhat singular that, if the Constitution were intended to be, by its framers and adopters, a slave-holding instrument, why neither slavery, slaveholding, nor slave can anywhere be found in it. What would be thought of an instrument, drawn up, legally drawn up, for the purpose of entitling the city of Rochester to a track of land, in which no mention of land was made? Now, there are certain rules of interpretation, for the proper understanding of all legal instruments.

These rules are well established. They are plain, common-sense rules, such as you and I, and all of us, can understand and apply, without having passed years in the study of law. I scout the idea that the question of the constitutionality or unconstitutionality of slavery is not a question for the people. I hold that every American citizen has a right to form an opinion of the constitution, and to propagate that opinion, and to use all honorable means to make his opinion the prevailing one.

Without this right, the liberty of an American citizen would be as insecure as that of a Frenchman. Ex-Vice-President Dallas tells us that the Constitution is an object to which no American mind can be too attentive, and no American heart too devoted. He further says, the Constitution, in its words, is plain and intelligible, and is meant for the home-bred, unsophisticated understandings of our fellow-citizens. Senator Berrien tell us that the Constitution is the fundamental law, that which controls all others. The charter of our liberties, which every citizen has a personal interest in understanding thoroughly. The testimony of Senator Breese, Lewis Cass, and many others that might be named, who are everywhere esteemed as sound lawyers, so regard the constitution. I take it, therefore, that it is not presumption in a private citizen to form an opinion of that instrument.

Now, take the Constitution according to its plain reading, and I defy the presentation of a single pro-slavery clause in it. On the other hand it will be found to contain principles and purposes, entirely hostile to the existence of slavery.

I have detained my audience entirely too long already. At some future period I will gladly avail myself of an opportunity to give this subject a full and fair discussion.

Allow me to say, in conclusion, notwithstanding the dark picture I have this day presented of the state of the nation, I do not despair of this country. There are forces in operation, which must inevitably work the downfall of slavery. "The arm of the Lord is not shortened," and the doom of slavery is certain. I, therefore, leave off where I began, with hope. While drawing encouragement from the Declaration of Independence, the great principles it contains, and the genius of American Institutions, my spirit is also cheered by the obvious tendencies of the age. Nations do not now stand in the same relation to each other that they did ages ago.

No nation can now shut itself up from the surrounding world, and trot round in the same old path of its fathers without interference. The time was when such could be done. Long established customs of hurtful character could formerly fence themselves in, and do their evil work with social impunity. Knowledge was then confined and enjoyed by the privileged few, and the multitude walked on in mental darkness. But a change has now come over the affairs of mankind. Walled cities and empires have become unfashionable. The arm of commerce has borne away the gates of the strong city. Intelligence is penetrating the darkest corners of the globe. It makes its pathway over and under the sea, as well as on the earth. Wind, steam, and lightning are its chartered agents. Oceans no longer divide, but link nations together.

From Boston to London is now a holiday excursion. Space is comparatively annihilated. Thoughts expressed on one side of the Atlantic, are distinctly heard on the other. The far off and almost fabulous Pacific rolls in grandeur at our feet. The Celestial Empire, the mystery of ages, is being solved. The fiat of the Almighty, "Let there be Light," has not yet spent its force. No abuse, no outrage whether in taste, sport or avarice, can now hide itself from the all-pervading light. The iron shoe, and crippled foot of China must be seen, in contrast with nature. Africa must rise and put on her yet unwoven garment. "Ethiopia shall stretch out her hand unto God." In the fervent aspirations of William Lloyd Garrison, I say, and let every heart join in saying it:

God speed the year of jubilee
The wide world o'er
When from their galling chains set free,
Th' oppress'd shall vilely bend the knee,

And wear the yoke of tyranny
Like brutes no more.
That year will come, and freedom's reign,

To man his plundered fights again
Restore.

God speed the day when human blood
Shall cease to flow!
In every clime be understood,
The claims of human brotherhood,
And each return for evil, good,
Not blow for blow;
That day will come all feuds to end.
And change into a faithful friend
Each foe.

God speed the hour, the glorious hour,
When none on earth
Shall exercise a lordly power,
Nor in a tyrant's presence cower;
But all to manhood's stature tower,
By equal birth!
That hour will come, to each, to all,
And from his prison-house, the thrall
Go forth.

Until that year, day, hour, arrive,
With head, and heart, and hand I'll strive,
To break the rod, and rend the gyve,
The spoiler of his prey deprive —
So witness Heaven!
And never from my chosen post,
Whate'er the peril or the cost,
Be driven.

A Glance at Ourselves — Conclusion,
by Martin R. Delany
(1852)

The Condition, Elevation, Emigration, and Destiny of the Colored People of the United States

With broken hopes — sad, devastation;

A race *resigned* to DEGREDATION!

We have said much to our young men and women, about their vocation and calling; we have dwelt much upon the menial position of our people in this county. Upon this point we cannot say too much, because there is a seeming satisfaction and seeking after such positions manifested on their part, unknown to any other people. There appears to be, a want of a sense of propriety or *self-respect*, altogether inexplicable because young men and women among us, many of whom have good trades and homes, adequate to their support, voluntarily leave them, and seek positions, such as servants, waiting maids, coachmen, nurses, cooks in gentlemens' kitchen, or such like occupations, when they can gain a livelihood at something more respectable, or elevating in character. And the worst part of the whole matter is that they have become so accustomed to it, it has become so "fashionable" that it seems to have become second nature, and they really become offended, when it is spoken against.

Among the German, Irish, and other European peasantry who came to this country, it matters not what they were employed at before and after they come; just so soon as they can better their condition by keeping shops, cultivating the soil, the young men and women going to night schools, qualifying themselves for usefulness, and learning trades, they do so. Their first and last care, object and aim is to better their condition by raising themselves above the condition that necessity places them in.

We do not say too much, when we say, as an evidence of the deep degradation of our race in the United States, that there are those among us, the wives and daughters, some of the *first ladies*, (and who dare say they are not the "first," because they belong to the "first class" and associate where any body among us can?) whose husbands are industrious, able and willing to support them, who voluntarily leave home, and become chambermaids, and stewardesses, upon vessels and steamboats, in all probability, to enable them to obtain some more fine or costly article of dress or furniture.

We have nothing to say against those whom *necessity* compels to do these things, those who can do no better; we have only to do with those who can, and will not, or do not do better. The whites are always in the advance, and we either standing still or retrograding; as that which does not go forward, must either stand in one place or go back. The father in all probability is a farmer, mechanic, or man of some independent business; and the wife, sons and daughters are chamber maids, on vessels, nurses and waiting maids, or coachmen and cooks in families. This is retrogradation. The wife, sons, and daughters should be elevated above this condition as a necessary consequence.

If we did not love our race superior to others, we would not concern ourself about their degradation; for the greatest desire of our heart is to see them stand on a level with the most elevated of mankind. No people are ever elevated above the condition of their *females*; hence the condition of the *mother* determines the condition of the child. To know the position of a people, it is only necessary to know the *condition* of their *females*; and despite themselves, they cannot rise above their level. Then what is our condition? Our *best ladies* being washerwomen, chamber maids, children's traveling nurses, and common house servants, and menials, we are all a degraded miserable people, inferior to any other people as a whole, on the face of the globe.

These great truths, however unpleasant, must be brought before the minds of our people in its true and proper light, as we have been too delicate about them, and too long concealed them for fear of giving offence. It would have been infinitely better for our race if these facts had been presented before us half a century ago, we would have been now proportionably benefitted by it.

As an evidence of the degradation to which we have been reduced, we dare premise, that this chapter will give offence to many, very many, and why? Because they may say, "He dared to say that the occupation of a *servant* is a degradation." It is not necessarily degrading; it would not be, to one or a few people of a kind; but a *whole race of servants* are a degradation to that people.

Efforts made by men of qualifications for the toiling and degraded millions among the whites, neither gives offence to that class, nor is it taken unkindly by them; but received with manifestations of gratitude; to know that they are thought to be equally worthy of, and entitled to stand on a level with the elevated classes; and they have only got to be informed of the way to raise themselves to make the effort and do so as far as they can.

But how different with us. Speak of our position in society, and it at once gives insult Though we are servants; among ourselves we claim to be *ladies* and *gentlemen*, equal in standing, and as the popular expression goes, "Just as good as any body" and so believing, we make no efforts to raise above the common level of menials, because the *best* being in that capacity, all are content with the position. We cannot at the same time, be domestic and lady; servant and gentleman.

We must be the one of the other. Sad, sad indeed, is the thought, that hangs drooping in our mind, when contemplating the picture drawn before us. Young men and women, "We write these things unto you, because ye are strong," because the writer, a few years ago, gave unpardonable offence to many of the young people of Philadelphia and other places, because he dared tell them, that he thought too much of them to be content with seeing them the servants of other people.

Surely, she that could be the mistress would not be the maid; neither would he that could be the master, be content with being the servant; then why be offended, when we point out to you, the way that leads from the menial to the mistress or the master. All this we seem to reject with fixed determination, repelling with anger, every effort on the part of our intelligent men and women to elevate us, with true Israelitish degradation, in reply to any suggestion or proposition that may be offered, "Who made thee a ruler and judge?"

The writer is no "Public Man," in the sense in which this is understood among our people, but simply an humble individual endeavoring to seek a livlihood by a profession obtained entirely by his own efforts, without relatives as he gained by the merit of his course and conduct, which he here gratefully acknowledges; and whatever he has accomplished, other young men may, by making corresponding efforts, also accomplish.

We have advised an emigration to Central and South America, and even to Mexico and the West Indies to those who prefer to either of the last named places, all of which are free countries, Brazil being the only real slave holding State in South America, there being nominal slavery in Dutch Guiana, Peru, Buenos Ayres, Paraguay, and Uruguay, in all of which places colored people have equality in social, civil, political, and religious privileges; Brazil making it punishable with death to import slaves into the empire.

Our oppressors, when urging us to go to Africa, tell us that we are better adapted to the climate than they, that the physical condition of the constitution of colored people better endures the heart of warm climates than that of the whites; this we are willing to *admit*, without argument, without adducing the physiological reason why, that colored people can and do stand warm climates better than whites; and find an answer fully to the point in the fact that they also stand modified that white people can stand; therefore, according to our oppressors's own showing, we are a *superior* race, being endowed with properties fitting us for *all parts* of the earth, while they are only adapted to *certain* parts. Of course, this proves our right and duty to live wherever we may *choose*; while the white race may only live where they *can*. We are content with the fact, and have ever claimed it. Upon this rock, they and we shall ever agree.

Of the West India Islands, Santa Cruz, belonging to Denmark; Porto Rico and Cuba with its little adjuncts, belonging to Spain, are the only slaveholding Islands among them, three fifths of the whole population of Cuba being colored people who cannot and will not much longer endure the burden and the yoke. They only want intelligent leaders of their own color, when they are ready at any moment to charge to the conflict to liberty or death.

The remembrance of the noble mulatoo, Placido, the gentleman, scholar, poet and intended Chief Engineer of the Army of Liberty and Freedom in Cuba; and the equally noble black, Charles Blair, who was to have been Commander In Chief, who were shamefully put to death in 1844, by that living monster, Captain General O'Donnell is still fresh and indelible to the mind of every bondsman of Cuba.

In our own country, the United States, there are *three millions five hundred thousand slaves*; and we, the nominally free colored people, are *six hundred thousand* in number; estimating one sixth to be men, we have *one hundred thousand* able bodied freeman, which will make a powerful auxiliary in any country to which we may become adopted, an ally not to be despised by any power on earth. We love our country, dearly love her, but she don't love us, she despises us and bids us begone, driving us from her embraces; but we do go, whatever love we have for her, we shall love the country none the less that receives us as her adopted children.

For the want of business habits and training, our energies have become paralyzed; our young men never think of business, anymore than if they were so many bondsmen, without the right to pursue any calling they may think most advisable. With our people in this country, dress and good appearances have been made the only test of gentleman and ladyship and that vocation which offers the best opportunity to dress and appear well has generally been preferred, however, menial and degrading by our young people, without even, in the majority of cases, an effort to do better; indeed, in many instances, refusing situations equally lucrative, and superior in position, but which would not allow as much display of dress and personal appearance. This, if we ever expect to rise, must be discarded from among us, and a high and respectable position assumed.

One of our great temporal curses is our consummate poverty. We are the poorest people as a class in the world of civilized mankind, abjectly, miserably poor, no one scarcely being able to assist the other. To this, of course, there are noble exceptions; but that which is common to, and the very process by which white men exist and succeed in life, is unknown to colored men in genera l. In any and every considerable community may be found, some one of our white fellow citizens, who is worth more than all the colored people in that community put together. We consequently have little or no efficiency. We must have men to be practically efficient in all the undertakings of life; and to obtain them, it is necessary that we should be engaged in lucrative pursuits, trade and general business transactions. In order to be thus engaged, it is necessary that we should occupy positions that afford the facilities for such pursuits.

To compete now with the mighty odds of wealth, social and religious preferences, and political influences of this country, at this advanced state of its national existence, we never may expect. A new country and new beginning is the only true rational, politic remedy for our disadvantageous position; and that country we have already pointed out, with triple golden advantages all things considered, to that of any country to which it has been the province of man to embark.

Every other than we have at various periods of necessity been a migratory people; and all when oppressed, shown a greater abhorrence of oppression, if not a greater love of liberty than we. We cling to our oppressors as the objects of our love. It is true that our enslaved brethren are here, and we have been led to believe that it is necessary for us to remain, on that account. Is it true, that all should remain in degradation, because a part are degraded? We believe no such thing. We believe it to be the duty of the Free to elevate themselves in the most speedy and effective manner possible; as the redemption of the bondman depends entirely upon the elevation of the freeman; therefore, to elevate the free colored people of America, anywhere upon this continent forebodes the speedy redemption of the slaves. We shall hope to hear no more of so fallacious a doctrine, the necessity of the free remaining in degradation for the sake of the oppressed. Let us apply, first, the lever to ourselves; and the force that elevates us to the position of manhood's considerations and honors will cleft the manacle of every slave in the land.

When such great worth and talents, for want of a better sphere, of men like Rev. Jonathan Robinson, Robert Douglass, Frederick A. Hinton, and a hundred others that might be named, were permitted to expire in a barber shop; and such living men as may be found in Boston, New York Philadelphia, Baltimore, Richmond, Washington City, Charleston, (S.C.), New Orleans, Cincinnati, Lousiville, St. Louis, Pittsburg, Buffalo, Rochester, Albany, Utica, Cleveland, Detroit, Milwaukee, Chicago, Columbus, Zanesville, Wheeling, and a hundred other places, confining themselves to Barber shops and waiterships in Hotels; certainly the necessity of such a course as we have pointed out must be cordially acknowledged; appreciated by every brother and sister of oppression; and not rejected as heretofore, as though they preferred inferiority to equality.

These minds must become "unfettered" and have "space to rise." This cannot be in their present positions. A continuance in any position becomes what is termed. "Second Nature;" it begins an *adaptation*, and *reconcilation* of *mind* to such condition. It changes the whole physiological condition of the system, and adapts man and woman to a higher or lower sphere in the pursuits of life. The offsprings of slaves and peasantry have the general characteristics of their parents; and nothing but a different course of training and education will change this character.

The slave may become a lover of his master, and learn to forgive him for continual deeds of maltreatment and abuse; just as the Spaniel would couch and fondle at the feet that kick him; because he has been taught to reverence them; and consequently becomes adapted in body and mind to his condition. Even the shrubbery loving Canary, and lofty soaring Eagle may be tamed to the cage, and learn to love it from habit of confinement. It has been so with us in our position among our oppressors; we have learned to love them.

When reflecting upon this all important, and to us, all absorbing subject; we feel in the agony and anxiety of the moment, as though we could cry out in the language of a Prophet of old: "Oh that my head were waters, and mine eyes a fountain of tears that I might weep day and night for the" degradation "of my people! Oh that I had in the wilderness a lodging place of wayfaring men; that I might leave my people and go from them"!

The Irishman and German in the United States are very different persons to what they were when in Ireland and Germany, the countries of their nativity. Their spirits were depressed and downcast; but the instant they set their foot upon unrestricted soil; free to act and untrammeled to move; their physical condition undergoes a change, which in time becomes physiological, which is transmitted to the offspring, who when born under the circumstances is a decidedly different being to what it would have been, had it been born under different circumstances.

A child born under oppression has all the elements of servility in its constitution; who when born under favorable circumstances, has to the contrary, all the elements of freedom and independence of feeling. Our children then may not be expected, to maintain that position and manly bearing; born under the unfavorable circumstances with which we are surrounded in this country; that we so much desire. To use the language of the talented Mr. Whipper, "they cannot be raised in this country without being stoop shouldered," Heaven's pathway stands unobstructed which will lead us into a Paradise of bliss. Let us go on and possess the land, and the God of Israel will be our God.

The lessons of every school book, the pages of every history, and columns of every newspaper are so replete with stimuli to nerve us on to manly aspirations that those of our young people, who will now refuse to enter upon this great theatre of Polynesian adventure, and take their position on the stage of Central and South America, where a brilliant engagement of certain and most triumphant success, in the drama of human equality awaits them; then with the blood of *slaves*, write upon the lintel of every door in sterling Capitals, to be gazed and hissed at by every passerby:

Doomed by the Creator, To servility and degradation; The SERVANT of the white man, And despised of every nation!

The Constitution of the United States: Is It Pro-Slavery or Anti-Slavery?,
by Frederick Douglass

Speech given in Glasgow, Scotland on March 26, 1860

I proceed to the discussion. And first a word about the question. Much will be gained at the outset if we fully and clearly understand the real question under discussion. Indeed, nothing is or can be understood. This are often confounded and treated as the same, for no better reason than that they resemble each other, even while they are in their nature and character totally distinct and even directly opposed to each other. This jumbling up things is a sort of dust-throwing which is often indulged in by small men who argue for victory rather than for truth. Thus, for instance, the American Government and the American Constitution are spoken of in a manner which would naturally lead the hearer to believe that one is identical with the other; when the truth is, they are distinct in character as is a ship and a compass.

The one may point right and the other steer wrong. A chart is one thing, the course of the vessel is another. The Constitution may be right, the Government is wrong. If the Government has been governed by mean, sordid, and wicked passions, it does not follow that the Constitution is mean, sordid, and wicked. What, then, is the question? I will state it. But first let me state what is not the question.

It is not whether slavery existed in the United States at the time of the adoption of the Constitution; it is not whether slaveholders took part in the framing of the Constitution; it is not whether those slaveholders, in their hearts, intended to secure certain advantages in that instrument for slavery; it is not whether the American Government has been wielded during seventy-two years in favour of the propagation and permanence of slavery; it is not whether a pro-slavery interpretation has been put upon the Constitution by the American Courts — all these points may be true or they may be false, they may be accepted or they may be rejected, without in any wise affecting the real question in debate.

The real and exact question between myself and the class of persons represented by the speech at the City Hall may be fairly stated thus: — 1st, Does the United States Constitution guarantee to any class or description of people in that country the right to enslave, or hold as property, any other class or description of people in that country? 2nd, Is the dissolution of the union between the slave and free States required by fidelity to the slaves, or by the just demands of conscience? Or, in other words, is the refusal to exercise the elective franchise, and to hold office in America, the surest, wisest, and best way to abolish slavery in America?

To these questions the Garrisonians say Yes. They hold the Constitution to be a slaveholding instrument, and will not cast a vote or hold office, and denounce all who vote or hold office, no matter how faithfully such persons labour to promote the abolition of slavery. I, on the other hand, deny that the Constitution guarantees the right to hold property in man, and believe that the way to abolish slavery in America is to vote such men into power as well use their powers for the abolition of slavery. This is the issue plainly stated, and you shall judge between us.

Before we examine into the disposition, tendency, and character of the Constitution, I think we had better ascertain what the Constitution itself is. Before looking for what it means, let us see what it is.

Here, too, there is much dust to be cleared away. What, then, is the Constitution? I will tell you. It is not even like the British Constitution, which is made up of enactments of Parliament, decisions of Courts, and the established usages of the Government.

The American Constitution is a written instrument full and complete in itself. No Court in America, no Congress, no President, can add a single word thereto, or take a single word threreto. It is a great national enactment done by the people, and can only be altered, amended, or added to by the people. I am careful to make this statement here; in America it would not be necessary. It would not be necessary here if my assailant had shown the same desire to be set before you the simple truth, which he manifested to make out a good case for himself and friends.

Again, it should be borne in mind that the mere text, and only the text, and not any commentaries or creeds written by those who wished to give the text a meaning apart from its plain reading, was adopted as the Constitution of the United States. It should also be borne in mind that the intentions of those who framed the Constitution, be they good or bad, for slavery or against slavery, are so respected so far, and so far only, as we find those intentions plainly stated in the Constitution. It would be the wildest of absurdities, and lead to endless confusion and mischiefs, if, instead of looking to the written paper itself, for its meaning, it were attempted to make us search it out, in the secret motives, and dishonest intentions, of some of the men who took part in writing it. It was what they said that was adopted by the people, not what they were ashamed or afraid to say, and really omitted to say.

Bear in mind, also, and the fact is an important one, that the framers of the Constitution sat with doors closed, and that this was done purposely, that nothing but the result of their labours should be seen, and that that result should be judged of by the people free from any of the bias shown in the debates. It should also be borne in mind, and the fact is still more important, that the debates in the convention that framed the Constitution, and by means of which a pro-slavery interpretation is now attempted to be forced upon that instrument, were not published till more than a quarter of a century after the presentation and the adoption of the Constitution.

These debates were purposely kept out of view, in order that the people should adopt, not the secret motives or unexpressed intentions of any body, but the simple text of the paper itself. Those debates form no part of the original agreement. I repeat, the paper itself, and only the paper itself, with its own plainly written purposes, is the Constitution. It must stand or fall, flourish or fade, on its own individual and self-declared character and objects. Again, where would be the advantage of a written Constitution, if, instead of seeking its meaning in its words, we had to seek them in the secret intentions of individuals who may have had something to do with writing the paper? What will the people of America a hundred years hence care about the intentions of the scriveners who wrote the Constitution?

These men are already gone from us, and in the course of nature were expected to go from us. They were for a generation, but the Constitution is for ages. Whatever we may owe to them, we certainly owe it to ourselves, and to mankind, and to God, to maintain the truth of our own language, and to allow no villainy, not even the villainy of holding men as slaves — which Wesley says is the sum of all villainies — to shelter itself under a fair-seeming and virtuous language. We owe it to ourselves to compel the devil to wear his own garments, and to make wicked laws speak out their wicked intentions. Common sense, and common justice, and sound rules of interpretation all drive us to the words of the law for the meaning of the law. The practice of the Government is dwelt upon with much fervour and eloquence as conclusive as to the slaveholding character of the Constitution. This is really the strong point and the only strong point, made in the speech in the City Hall. But good as this argument is, it is not conclusive.

A wise man has said that few people have been found better than their laws, but many have been found worse. To this last rule America is no exception. Her laws are one thing, her practice is another thing. We read that the Jews made void the law by their tradition, that Moses permitted men to put away their wives because of the hardness of their hearts, but that this was not so at the beginning. While good laws will always be found where good practice prevails, the reverse does not always hold true. Far from it. The very opposite is often the case.

What then? Shall we condemn the righteous law because wicked men twist it to the support of wickedness? Is that the way to deal with good and evil? Shall we blot out all distinction between them, and hand over to slavery all that slavery may claim on the score of long practice? Such is the course commended to us in the City Hall speech. After all, the fact that men go out of the Constitution to prove it pro-slavery, whether that going out is to the practice of the Government, or to the secret intentions of the writers of the paper, the fact that they do go out is very significant. It is a powerful argument on my side. It is an admission that the thing for which they are looking is not to be found where only it ought to be found, and that is in the Constitution itself. If it is not there, it is nothing to the purpose, be it wheresoever else it may be. But I shall have no more to say on this point hereafter.

The very eloquent lecturer at the City Hall doubtless felt some embarrassment from the fact that he had literally to *give* the Constitution a pro-slavery interpretation; because upon its face it of itself conveys no such meaning, but a very opposite meaning. He thus sums up what he calls the slaveholding provisions of the Constitution. I quote his own words: — "Article 1, section 9, provides for the continuance of the African slave trade for the 20 years, after the adoption of the Constitution. Art. 4, section 9, provides for the recovery from the other States of fugitive slaves. Art. 1, section 2, gives the slave States a representation of the three-fifths of all the slave population; and Art. 1, section 8, requires the President to use the military, naval, ordnance, and militia resources of the entire country for the suppression of slave insurrection, in the same manner as he would employ them to repel invasion."

Now any man reading this statement, or hearing it made with such a show of exactness, would unquestionably suppose that he speaker or writer had given the plain written text of the Constitution itself. I can hardly believe that the intended to make any such impression. It would be a scandalous imputation to say he did. Any yet what are we to make of it? How can we regard it? How can he be screened from the charge of having perpetrated a deliberate and point-blank misrepresentation?

That individual has seen fit to place himself before the public as my opponent, and yet I would gladly find some excuse for him. I do not wish to think as badly of him as this trick of his would naturally lead me to think. Why did he not read the Constitution? Why did he read that which was not the Constitution? He pretended to be giving chapter and verse, section and clause, paragraph and provision. The words of the Constitution were before him. Why then did he not give you the plain words of the Constitution?

Oh, sir, I fear that the gentleman knows too well why he did not. It so happens that no such words as "African slave trade," no such words as "slave insurrections," are anywhere used in that instrument. These are the words of that orator, and not the words of the Constitution of the United States.

Now you shall see a slight difference between my manner of treating this subject and what which my opponent has seen fit, for reasons satisfactory to himself, to pursue. What he withheld, that I will spread before you: what he suppressed, I will bring to light: and what he passed over in silence, I will proclaim: that you may have the whole case before you, and not be left to depend upon either his, or upon my inferences or testimony.

Here then are several provisions of the Constitution to which reference has been made. I read them word for word just as they stand in the paper, called the United States Constitution, Art. I, sec. 2. "Representatives and direct taxes shall be apportioned among the several States which may be included in this Union, according to their respective numbers, which shall be determined by adding to the whole number of free persons, including those bound to service for a term years, and excluding Indians not taxed, three-fifths of all other persons; Art. I, sec. 9. The migration or importation of such persons as any of the States now existing shall think fit to admit, shall not be prohibited by the Congress prior to the year one thousand eight hundred and eight, but a tax or duty may be imposed on such importation, not exceeding tend dollars for each person; Art. 4, sec. 2. No person held to service or labour in one State, under the laws thereof, escaping into another shall, in consequence of any law or regulation therein, be discharged from service or labour; but shall be delivered up on claim of the party to whom such service or labour may be due; Art. I, sec. 8.

To provide for calling for the militia to execute the laws of the Union, suppress insurrections, and repel invasions." Here then, are those provisions of the Constitution, which the most extravagant defenders of slavery can claim to guarantee a right of property in man. These are the provisions which have been pressed into the service of the human fleshmongers of America. Let us look at them just as they stand, one by one. Let us grant, for the sake of the argument, that the first of these provisions, referring to the basis of representation and taxation, does refer to slaves. We are not compelled to make that admission, for it might fairly apply to aliens — persons living in the country, but not naturalized. But giving the provisions the very worse construction, what does it amount to?

I answer — It is a downright disability laid upon the slaveholding States; one which deprives those States of two-fifths of their natural basis of representation. A black man in a free State is worth just two-fifths more than a black man in a slave State, as a basis of political power under the Constitution. Therefore, instead of encouraging slavery, the Constitution encourages freedom by giving an increase of "two-fifths" of political power to free over slave States. So much for the three-fifths clause; taking it at is worst, it still leans to freedom, not slavery; for, be it remembered that the Constitution nowhere forbids a coloured man to vote. I come to the next, that which it is said guaranteed the continuance of the African slave trade for twenty years. I will also take that for just what my opponent alleges it to have been, although the Constitution does not warrant any such conclusion.

But, to be liberal, let us suppose it did, and what follows? Why, this — that this part of the Constitution, so far as the slave trade is concerned, became a dead letter more than 50 years ago, and now binds no man's conscience for the continuance of any slave trade whatsoever. Mr. Thompson is just 52 years too late in dissolving the Union on account of this clause. He might as well dissolve the British Government, because Queen Elizabeth granted to Sir John Hawkins to import Africans into the West Indies 300 years ago! But there is still more to be said about this abolition of the slave trade. Men, at that time, both in England and in America, looked upon the slave trade as the life of slavery. The abolition of the slave trade was supposed to be the certain death of slavery. Cut off the stream, and the pond will dry up, was the common notion at the time.

Wilberforce and Clarkson, clear-sighted as they were, took this view; and the American statesmen, in providing for the abolition of the slave trade, thought they were providing for the abolition of the slavery. This view is quite consistent with the history of the times. All regarded slavery as an expiring and doomed system, destined to speedily disappear from the country. But, again, it should be remembered that this very provision, if made to refer to the African slave trade at all, makes the Constitution anti-slavery rather than for slavery; for it says to the slave States, the price you will have to pay for coming into the American Union is, that the slave trade, which you would carry on indefinitely out of the Union, shall be put an end to in twenty years if you come into the Union. Secondly, if it does apply, it expired by its own limitation more than fifty years ago.

Thirdly, it is anti-slavery, because it looked to the abolition of slavery rather than to its perpetuity. Fourthly, it showed that the intentions of the framers of the Constitution were good, not bad. I think this is quite enough for this point. I go to the "slave insurrection" clause, though, in truth, there is no such clause. The one which is called so has nothing whatever to do with slaves or slaveholders any more than your laws for suppression of popular outbreaks has to do with making slaves of you and your children. It is only a law for suppression of riots or insurrections.

But I will be generous here, as well as elsewhere, and grant that it applies to slave insurrections. Let us suppose that an anti-slavery man is President of the United States (and the day that shall see this the case is not distant) and this very power of suppressing slave insurrections would put an end to slavery. The right to put down an insurrection carries with it the right to determine the means by which it shall be put down. If it should turn out that slavery is a source of insurrection, that there is no security from insurrection while slavery lasts, why, the Constitution would be best obeyed by putting an end to slavery, and an anti-slavery Congress would do the very same thing.

Thus, you see, the so-called slave-holding provisions of the American Constitution, which a little while ago looked so formidable, are, after all, no defence or guarantee for slavery whatever. But there is one other provision. This is called the "Fugitive Slave Provision." It is called so by those who wish to make it subserve the interest of slavery in America, and the same by those who wish to uphold the views of a party in this country. It is put thus in the speech at the City Hall: — "Let us go back to 1787, and enter Liberty Hall, Philadelphia, where sat in convention the illustrious men who framed the Constitution — with George Washington in the chair.

On the 27th of September, Mr. Butler and Mr. Pinckney, two delegates from the State of South Carolina, moved that the Constitution should require that fugitive slaves and servants should be delivered up like criminals, and after a discussion on the subject, the clause, as it stands in the Constitution, was adopted. After this, in the conventions held in the several States to ratify the Constitution, the same meaning was attached to the words. For example, Mr. Madison (afterwards President), when recommending the Constitution to his constituents, told them that the clause would secure them their property in slaves." I must ask you to look well to this statement. Upon its face, it would seem a full and fair statement of the history of the transaction it professes to describe and yet I declare unto you, knowing as I do the facts in the case, my utter amazement at the downright untruth conveyed under the fair seeming words now quoted.

The man who could make such a statement may have all the craftiness of a lawyer, but who can accord to him the candour of an honest debater? What could more completely destroy all confidence in his statements? Mark you, the orator had not allowed his audience to hear read the provision of the Constitution to which he referred. He merely characterized it as one to "deliver up fugitive slaves and servants like criminals," and tells you that this was done "after discussion." But he took good care not to tell you what was the nature of that discussion. He have would have spoiled the whole effect of his statement had he told you the whole truth.

Now, what are the facts connected with this provision of the Constitution? You shall have them. It seems to take two men to tell the truth. It is quite true that Mr. Butler and Mr. Pinckney introduced a provision expressly with a view to the recapture of fugitive slaves: it is quite true also that there was some discussion on the subject — and just here the truth shall come out. These illustrious kidnappers were told promptly in that discussion that no such idea as property in man should be admitted into the Constitution. The speaker in question might have told you, and he would have told you but the simple truth, if he had told you that he proposition of Mr. Butler and Mr. Pinckney — which he leads you to infer was adopted by the convention that from the Constitution — was, in fact, promptly and indignantly rejected by that convention.

He might have told you, had it suited his purpose to do so, that the words employed in the first draft of the fugitive slave clause were such as applied to the condition of slaves, and expressly declared that persons held to "servitude" should be given up; but that the word "servitude" was struck from the provision, for the very reason that it applied to slaves. He might have told you that the same Mr. Madison declared that the word was struck out because the convention would not consent that the idea of property in men should be admitted into the Constitution.

The fact that Mr. Madison can be cited on both sides of this question is another evidence of the folly and absurdity of making the secret intentions of the framers the criterion by which the Constitution is to be construed. But it may be asked — if this clause does not apply to slaves, to whom does it apply?

I answer, that when adopted, it applies to a very large class of persons — namely, redemptioners — persons who had come to America from Holland, from Ireland, and other quarters of the globe — like the Coolies to the West Indies — and had, for a consideration duly paid, become bound to "serve and labour" for the parties two whom their service and labour was due. It applies to indentured apprentices and others who have become bound for a consideration, under contract duly made, to serve and labour, to such persons this provision applies, and only to such persons. The plain reading of this provision shows that it applies, and that it can only properly and legally apply, to persons "bound to service." Its object plainly is, to secure the fulfillment of contracts for "service and labour." It applies to indentured apprentices, and any other persons from whom service and labour may be due.

The legal condition of the slave puts him beyond the operation of this provision. He is not described in it. He is a simple article of property. He does not owe and cannot owe service. He cannot even make a contract. It is impossible for him to do so. He can no more make such a contract than a horse or an ox can make one. This provision, then, only respects persons who owe service, and they only can owe service who can receive an equivalent and make a bargain. The slave cannot do that, and is therefore exempted from the operation of this fugitive provision. In all matters where laws are taught to be made the means of oppression, cruelty, and wickedness, I am for strict construction. I will concede nothing.

It must be shown that it is so nominated in the bond. The pound of flesh, but not one drop of blood. The very nature of law is opposed to all such wickedness, and makes it difficult to accomplish such objects under the forms of law. Law is not merely an arbitrary enactment with regard to justice, reason, or humanity. Blackstone defines it to be a rule prescribed by the supreme power of the State commanding what is right and forbidding what is wrong. The speaker at the City Hall laid down some rules of legal interpretation. These rules send us to the history of the law for its meaning. I have no objection to such a course in ordinary cases of doubt. But where human liberty and justice are at stake, the case falls under an entirely different class of rules. There must be something more than history — something more than tradition. The Supreme Court of the United States lays down this rule, and it meets the case exactly — "Where rights are infringed — where the fundamental principles of the law are overthrown — where the general system of the law is departed from, the legislative intention must be expressed with irresistible clearness."

The same court says that the language of the law must be construed strictly in favour of justice and liberty. Again, there is another rule of law. It is — Where a law is susceptible of two meanings, the one making it accomplish an innocent purpose, and the other making it accomplish a wicked purpose, we must in all cases adopt that which makes it accomplish an innocent purpose. Again, the details of a law are to be interpreted in the light of the declared objects sought by the law. I set these rules down against those employed at the City Hall. To me they seem just and rational.

I only ask you to look at the American Constitution in the light of them, and you will see with me that no man is guaranteed a right of property in man, under the provisions of that instrument. If there are two ideas more distinct in their character and essence than another, those ideas are "persons" and "property," "men" and "things." Now, when it is proposed to transform persons into "property" and men into beasts of burden, I demand that the law that completes such a purpose shall be expressed with irresistible clearness.

The thing must not be left to inference, but must be done in plain English. I know how this view of the subject is treated by the class represented at the City Hall. They are in the habit of treating the Negro as an exception to general rules. When their own liberty is in question they will avail themselves of all rules of law which protect and defend their freedom; but when the black man's rights are in question they concede everything, admit everything for slavery, and put liberty to the proof. They reserve the common law usage, and presume the Negro a slave unless he can prove himself free.

I, on the other hand, presume him free unless he is proved to be otherwise. Let us look at the objects for which the Constitution was framed and adopted, and see if slavery is one of them. Here are its own objects as set forth by itself: — "We, the people of these United States, in order to form a more perfect union, establish justice, ensure domestic tranquility, provide for the common defense, promote the general welfare, and secure the blessings of liberty to ourselves and our posterity, do ordain and establish this Constitution of the United States of America."

The objects here set forth are six in number: union, defence, welfare, tranquility, justice, and liberty. These are all good objects, and slavery, so far from being among them, is a foe of them all. But it has been said that Negroes are not included within the benefits sought under this declaration. This is said by the slaveholders in America — it is said by the City Hall orator — but it is not said by the Constitution itself.

Its language is "we the people;" not we the white people, not even we the citizens, not we the privileged class, not we the high, not we the low, but we the people; not we the horses, sheep, and swine, and wheel-barrows, but we the people, we the human inhabitants; and, if Negroes are people, they are included in the benefits for which the Constitution of America was ordained and established. But how dare any man who pretends to be a friend to the Negro thus gratuitously concede away what the Negro has a right to claim under the Constitution? Why should such friends invent new arguments to increase the hopelessness of his bondage?

This, I undertake to say, as the conclusion of the whole matter, that the constitutionality of slavery can be made out only by disregarding the plain and common-sense reading of the Constitution itself; by discrediting and casting away as worthless the most beneficent rules of legal interpretation; by ruling the Negro outside of these beneficent rules; by claiming that the Constitution does not mean what it says, and that it says what it does not mean; by disregarding the written Constitution, and interpreting it in the light of a secret understanding. It is in this mean, contemptible, and underhand method that the American Constitution is pressed into the service of slavery.

They go everywhere else for proof that the Constitution declares that no person shall be deprived of life, liberty, or property without due process of law; it secures to every man the right of trial by jury, the privilege of the writ of habeas corpus — the great writ that put an end to slavery and slave-hunting in England — and it secures to every State a republican form of government. Anyone of these provisions in the hands of abolition statesmen, and backed up by a right moral sentiment, would put an end to slavery in America. The Constitution forbids the passing of a bill of attainder: that is, a law entailing upon the child the disabilities and hardships imposed upon the parent.

Every slave law in America might be repealed on this very ground. The slave is made a slave because his mother is a slave. But to all this it is said that the practice of the American people is against my view. I admit it. They have given the Constitution a slaveholding interpretation. I admit it. Thy have committed innumerable wrongs against the Negro in the name of the Constitution. Yes, I admit it all; and I go with him who goes farthest in denouncing these wrongs. But it does not follow that the Constitution is in favour of these wrongs because the slaveholders have given it that interpretation. To be consistent in his logic, the City Hall speaker must follow the example of some of his brothers in America — he must not only fling away the Constitution, but the Bible.

The Bible must follow the Constitution, for that, too, has been interpreted for slavery by American divines. Nay, more, he must not stop with the Constitution of America, but make war with the British Constitution, for, if I mistake not, the gentleman is opposed to the union of Church and State. In America he called himself a Republican. Yet he does not go for breaking down the British Constitution, although you have a Queen on the throne, and bishops in the House of Lords.

My argument against the dissolution of the American Union is this: It would place the slave system more exclusively under the control of the slaveholding States, and withdraw it from the power in the Northern States which is opposed to slavery. Slavery is essentially barbarous in its character. It, above all things else, dreads the presence of an advanced civilisation. It flourishes best where it meets no reproving frowns, and hears no condemning voices.

While in the Union it will meet with both. Its hope of life, in the last resort, is to get out of the Union. I am, therefore, for drawing the bond of the Union more completely under the power of the Free States. What they most dread, that I most desire. I have much confidence in the instincts of the slaveholders. They see that the Constitution will afford slavery no protection when it shall cease to be administered by slaveholders.

They see, moreover, that if there is once a will in the people of America to abolish slavery, this is no word, no syllable in the Constitution to forbid that result. They see that the Constitution has not saved slavery in Rhode Island, in Connecticut, in New York, or Pennsylvania; that the Free States have only added three to their original number. There were twelve Slave States at the beginning of the Government: there are fifteen now. They dissolution of the Union would not give the North a single advantage over slavery, but would take from it many. Within the Union we have a firm basis of opposition to slavery. It is opposed to all the great objects of the Constitution. The dissolution of the Union is not only an unwise but a cowardly measure — 15 millions running away from three hundred and fifty thousand slaveholders.

Mr. Garrison and his friends tell us that while in the Union we are responsible for slavery. He and they sing out "No Union with slaveholders," and refuse to vote. I admit our responsibility for slavery while in the Union but I deny that going out of the Union would free us from that responsibility. There now clearly is no freedom from responsibility for slavery to any American citizen short to the abolition of slavery.

The American people have gone quite too far in this slaveholding business now to sum up their whole business of slavery by singing out the cant phrase, "No union with slaveholders." To desert the family hearth may place the recreant husband out of the presence of his starving children, but this does not free him from responsibility. If a man were on board of a pirate ship, and in company with others had robbed and plundered, his whole duty would not be preformed simply by taking the longboat and singing out, "No union with pirates." His duty would be to restore the stolen property.

The American people in the Northern States have helped to enslave the black people. Their duty will not have been done till they give them back their plundered rights. Reference was made at the City Hall to my having once held other opinions, and very different opinions to those I have now expressed. An old speech of mine delivered fourteen years ago was read to show — I know not what. Perhaps it was to show that I am not infallible. If so, I have to say in defence, that I never pretended to be.

Although I cannot accuse myself of being remarkably unstable, I do not pretend that I have never altered my opinion both in respect to men and things. Indeed, I have been very much modified both in feeling and opinion within the last fourteen years. When I escaped from slavery, and was introduced to the Garrisonians, I adopted very many of their opinions, and defended them just as long as I deemed them true. I was young, had read but little, and naturally took some things on trust. Subsequent experience and reading have led me to examine for myself. This had brought me to other conclusions. When I was a child, I thought and spoke as a child. But the question is not as to what were my opinions fourteen years ago, but what they are now. If I am right now, it really does not matter what I was fourteen years ago.

My position now is one of reform, not of revolution. I would act for the abolition of slavery through the Government — not over its ruins. If slaveholders have ruled the American Government for the last fifty years, let the anti-slavery men rule the nation for the next fifty years. If the South has made the Constitution bend to the purposes of slavery, let the North now make that instrument bend to the cause of freedom and justice. If 350,000 slaveholders have, by devoting their energies to that single end, been able to make slavery the vital and animating spirit of the American Confederacy for the last 72 years, now let the freemen of the North, who have the power in their own hands, and who can make the American Government just what they think fit, resolve to blot out for ever the foul and haggard crime, which is the blight and mildew, the curse and the disgrace of the whole United States.

The Destiny of Colored Americans,
by Frederick Douglass

The North Star

November 16, 1849

It is impossible to settle, by the light of the present, and by the experience of the past, anything, definitely and absolutely, as to the future condition of the colored people of this country; but, so far as present indications determine, it is clear that this land must continue to be the home of the colored man so long as it remains the abode of civilization and religion. For more than two hundred years we have been identified with its soil, its products, and its institutions; under the sternest and bitterest circumstances of slavery and oppression — under the lash of Slavery at the South — under the sting of prejudice and malice at the North — and under hardships the most unfavorable to existence and population, we have lived, and continue to live and increase. The persecuted red man of the forest, the original owner of the soil, has, step by step, retreated from the Atlantic lakes and rivers; escaping, as it were, before the footsteps of the white man, and gradually disappearing from the face of the country.

He looks upon the steamboats, the railroads, and canals, cutting and crossing his former hunting grounds; and upon the ploughshare, throwing up the bones of his venerable ancestors, and beholds his glory departing — and his heart sickens at the desolation. He spurns the civilization — he hates the race which has despoiled him, and unable to measure arms with his superior foe, he dies.

Not so with the black man. More unlike the European in form, feature and color — called to endure greater hardships, injuries and insults than those to which the Indians have been subjected, he yet lives and prospers under every disadvantage. Long have his enemies sought to expatriate him, and to teach his children that this is not their home, but in spite of all their cunning schemes, and subtle contrivances, his footprints yet mark the soil of his birth, and he gives every indication that America will, forever, remain the home of his posterity. We deem it a settled point that the destiny of the colored man is bound up with that of the white people of his country; be the destiny of the latter what it may.

It is idle — worse than idle, ever to think of our expatriation, or removal. The history of the colonization society must extinguish all such speculations. We are rapidly filling up the number of four millions; and all the gold of California combined, would be insufficient to defray the expenses attending our colonization. We are, as laborers, too essential to the interests of our white fellow-countrymen, to make a very grand effort to drive us from this country among probable events. While labor is needed, the labor cannot fail to be valued; and although passion and prejudice may sometimes vociferate against us, and demand our expulsion, such efforts will only be spasmodic, and can never prevail against the sober second thought of self-interest. *We are here*, and here we are likely to be. To imagine that we shall ever be eradicated is absurd and ridiculous. We can be remodified, changed, and assimilated, but never extinguished.

We repeat, therefore, that *we are here*; and that this is *our* country; and the question for the philosophers and statesmen of the land ought to be, What principles should dictate the policy of the action towards us? We shall neither die out, nor be driven out; but shall go on with this people, either as a testimony against them, or as an evidence in their favor throughout their generations. We are clearly on their hands, and must remain there forever. All this we say for the benefit of those who hate the Negro more than they love their country. In an article, under the caption of "Government and its Subjects," (published in our last week's paper,) we called attention to the unwise, as well as the unjust policy usually adopted, by our Government, towards its colored citizens. We would continue to direct our attention to that policy, and in our humble way, we would remonstrate against it, as fraught with evil to the white man, as well as to his victim.

The white man's happiness cannot be purchased by the black man's misery. Virtue cannot prevail among the white people, by its destruction among the black people, who form a part of the whole community. It is evident that the white and black "must fall or flourish together." In the light of this great truth, laws ought to be enacted, and institutions established — all distinctions, founded on complexion, ought to be repealed, repudiated, and forever abolished — and every right, privilege, and immunity, now enjoyed by the white man, ought to be as freely granted to the man of color.

Where "knowledge is power," that nation is the most powerful which has the largest population of intelligent men; for a nation to cramp, and circumscribe the mental faculties of a class of its inhabitants, is as unwise as it is cruel, since it, in the same proportion, sacrifices its power and happiness. The American people, in the light of this reasoning, are at this moment, in obedience to their pride and folly, (we say nothing of the wickedness of the act,) wasting one sixth part of the energies of the entire nation by transforming three millions of its men into beasts of burden. — What a loss to industry, skill, invention, (to say nothing of its foul and corrupting influence,) is *Slavery*!

How it ties the hand, cramps the mind, darkens the understanding, and paralyses the whole man! Nothing is more evident to a man who reasons at all, than that America is acting an irrational part in continuing the slave system at the South, and in oppressing its free colored citizens at the North. Regarding the nation as an individual, the act of enslaving and oppressing thus, is as wild and senseless as it would be for Nicholas to order the amputation of the right arm of every Russian soldier before engaging in war with France. We again repeat that Slavery is the peculiar weakness of America, as well as its peculiar crime; and the day may yet come with this visionary and oft repeated declaration will be found to contain a great truth.

Emancipation Proclamation

Abraham Lincoln

January 1, 1863

By the President of the United States of America:
A Proclamation.

Whereas, on the twenty-second day of September, in the year of our Lord one thousand eight hundred and sixty two, a proclamation was issued by the President of the United States, containing, among other things, the following, to wit:

"That on the first day of January, in the year of our Lord one thousand eight hundred and sixty-three, all persons held as slaves within any State or designated part of a State, the people whereof shall then be in rebellion against the United States, shall be then, thenceforward, and forever free; and the Executive Government of the United States, including the military and naval authority thereof, will recognize and maintain the freedom of such persons, and will do no act or acts to repress such persons, or any of them, in any efforts they may make for their actual freedom.

"That the Executive will, on the first day of January aforesaid, by proclamation, designate the States and parts of States, if any, in which the people thereof, respectively, shall then be in rebellion against the United States; and the fact that any State, or the people thereof, shall on that day be, in good faith, represented in the Congress of the United States by members chosen thereto at elections wherein a majority of the qualified voters of such State shall have participated, shall, in the absence of strong countervailing testimony, be deemed conclusive evidence that such State, and the people thereof, are not then in rebellion against the United States."

Now, therefore I, Abraham Lincoln, President of the United States, by virtue of the power in me vested as Commander-in-Chief, of the Army and Navy of the United States in time of actual armed rebellion against authority and government of the United States, and as a fit and necessary war measure for suppressing said rebellion, do, on this first day of January, in the year of our Lord one thousand eight hundred and sixty three, and in accordance with my purpose so to do publicly proclaimed for the full period of one hundred days, from the day first above mentioned, order and designate as the States and parts of States wherein the people thereof respectively, are this day in rebellion against the United States, the following, to wit:

Arkansas, Texas, Louisiana, (except the Parishes of St. Bernard, Plaquemines, Jefferson, St. Johns, St. Charles, St. James, Ascension, Assumption, Terrebonne, Lafourche, St. Mary, St. Martin, and Orleans, including the City of New-Orleans) Mississippi, Alabama, Florida, Georgia, South-Carolina, North-Carolina, and Virginia, (except the forty-eight counties designated as West Virginia, and also the counties of Berkley, Accomac, Northampton, Elizabeth-City, York, Princess Ann, and Norfolk, including the cities of Norfolk & Portsmouth); and which excepted parts are, for the present, left precisely as if this proclamation were not issued.

And by virtue of the power, and for the purpose aforesaid, I do order and declare that all persons held as slaves within said designated States, and parts of States, are, and henceforward shall be free; and that the Executive government of the United States, including the military and naval authorities thereof, will recognize and maintain the freedom of said persons.

And I hereby enjoin upon the people so declared to be free to abstain from all violence, unless in necessary self-defense; and I recommend to them that, in all cases when allowed, they labor faithfully for reasonable wages.

And I further declare and make known, that such persons of suitable condition, will be received into the armed service of the United States to garrison forts, positions, stations, and other places, and to man vessels of all sorts in said service.

And upon this act, sincerely believed to be an act of justice, warranted by the Constitution, upon military necessity, I invoke the considerate judgment of mankind, and the gracious favor of Almighty God.

In witness whereof, I have hereunto set my hand and caused the seal of the United States to be affixed.

Done at the City of Washington, this first day of January, in the year of our Lord one thousand eight hundred and sixty three, and of the Independence of the United States of America the eighty-seventh.

By the President: ABRAHAM LINCOLN

WILLIAM H. SEWARD, Secretary of State.

President Abraham Lincoln- Alexander Gardner (LOC)

Why Should a Colored Man Enlist?
by Frederick Douglass
(1863)

This question has been repeatedly put to us while raising men for the 54th Massachusetts regiment during the past five weeks, and perhaps we cannot at present do a better service to the cause of our people or to the cause of the country than by giving a few of the many reasons why a colored man should enlist.

First. You are a man, although a colored man. If you were only a horse or an ox, incapable of deciding whether the rebels are right or wrong, you would have no responsibility, and might like a horse or an ox go on eating your corn or grass, in total indifference, as to which side is victorious or vanquished in this conflict. You are however not horse, and no ox, but a man, and whatever concerns man should interest you. He who looks upon a conflict between right and wrong, and does not help the right against the wrong, despises and insults his own nature, and invites the contempt of mankind. As between the North and South, the North is clearly on the right side and the South is flagrantly in the wrong.

You should therefore, simply as a matter of right and wrong, give your utmost aid to the North. In presence of such a contest there is not neutrality for any man. You are either for the Government or against the Government. Manhood requires you to take sides, and you are mean or noble according to how you choose between action and inaction. — If you are sound in body and mind, there is nothing in your *color* to excuse you from enlisting in the service of the republic against its enemies. If *color* should not be a criterion of rights, neither should it be a standard of duty. The whole duty of a man, belongs alike to white and black.

"*A man's a man for a' that.*" Second. You are however, not only a man, but an American citizen, so declared by the highest legal advisor of the Government, and you have hitherto expressed in various ways, not only your willingness but your earnest desire to fulfill any and every obligation which the relation of citizenship imposes. Indeed, you have hitherto felt wronged and slighted, because while white men of all other nations have been freely enrolled to serve the country, you are a native-born citizen and have been coldly denied the honor of aiding in defense of the land of your birth.

The injustice thus done to you is now repented of by the Government and you are welcomed to a place in the army of the nation. Should you refuse to enlist *now*, you will justify the past contempt of the Government towards you and lead it to regret having honored you with a call to take up arms in its defense. You cannot but see that here is a good reason why you should promptly enlist.

Third. A third reason why a colored man should enlist is found in the fact that every Negro-hated and slavery-lover in the land regards the arming of Negroes as a calamity and is doing its best to prevent it. Even now all the weapons of malice, in the shape of slander and ridicule, are used to defeat the filling up of the 54th Massachusetts (colored) regiment. In nine cases out of ten, you will find it safe to do just what your enemy would gladly have you leave undone. What helps you hurts him. Find out what he does not want and give him a plenty of it.

Fourth. You should enlist to learn the use of arms, to become familiar with the means of securing, protecting and defending your own liberty. A day may come when men shall learn war no more, when justice shall be so clearly apprehended, so universally practiced, and humanity shall be so profoundly loved and respected, that war and bloodshed shall be confined only to beasts of prey. Manifestly however, that time has not yet come, and while all men should labor to hasten its coming, by the cultivation of all the elements conducive to peace, it is plain that for the present no race of men can depend wholly upon moral means the maintenance of their rights. Men must either be governed by love or by fear.

They must love to do right or fear to do wrong. The only way open to any race to make their rights respected is to learn how to defend them. When it is seen that black men no more than white men can be enslaved with impunity, men will be less inclined to enslave and oppress them. Enlist, therefore, that you may learn the art and assert the ability to defend yourself and your race.

Fifth. You are a member of a long enslaved and despised race. Men have set down your submission to Slavery and insult, to a lack of manly courage. They point to this fact as demonstrating your fitness only to be a servile class. You should enlist and disprove the slander, and wipe out the reproach. When you shall be seen nobly defending the liberties of you own country against rebels and traitors — brass itself will blush to use such arguments imputing cowardice against you.

Sixth. Whether you are or are not, entitled to all the rights of citizenship in this country has long been a matter of dispute to your prejudice. By enlisting in the service of your country at this trial hour, and upholding the National Flag, you stop the mouths of traducers and win applause even from the lips of ingratitude. Enlist and you will make this your country in common with all other men born in the country or out of it.

Seventh. Enlist for your own sake. Decried and decried as you have been and still are, you need an act of this kind by which to recover your own self-respect. You have to some extent rated your value by the estimate of your enemies and hence have counted yourself less than you are. You owe it to yourself and your race to rise from your social debasement and take your place among the soldiers of your country, a man among them. Depend upon it, the subjective effect of this one act of enlisting will be immense and highly beneficial. You will stand more erect, walk more assured, feel more at ease, and be less liable to insult than you ever were before. He who fights and battle of America may claim America as his own country — and have that claim respected. Thus in defending your country now against rebels and traitors you are defending your own liberty, honor, manhood and self-respect.

Eighth. You should enlist because your doing so will be one of the most certain means of preventing the country from drifting back into the whirlpool of Pro-Slavery Compromise at the end of the war, which is now our greatest danger. He who shall witness another Compromise with Slavery in this country will see the free colored man of the North more than ever a victim of the pride, lust, scorn and violence of all classes of white men. The whole North will be but another Detroit, where every white fiend may with impunity revel in unrestrained beastliness towards people of color; they may burn their houses, insult their wives and daughters, and kill indiscriminately. If you mean to live in this country now is the time for you to do your full share in making it a country where you can your children after you can live in comparative safety. Prevent a compromise with the traitors, compel them to come back to the Union whipped and humbled into obedience and all will be well. But let them come back as masters and all their hate and hellish ingenuity will be exerted to stir up the ignorant masses of the North to hate, hinder and persecute the free colored people of the North. That most inhuman of all modern enactments, with its bribed judges, and summary process, the Fugitive

Slave Law, with all its infernal train of canting divines, preaching the gospel of kidnapping, as twelve years ago, will be revived against the free colored people of the North. One or two black brigades will do much to prevent all this.

Ninth. You should enlist because the war for the Union, whether men so call it or not, is a war for Emancipation. The salvation of the country, by the inexorable relation of cause and effect, can be secured only by the complete abolition of Slavery. The President has already proclaimed emancipation to the Slaves in rebel States which is tantamount to declaring Emancipation in all the States, for Slavery must exist everywhere in the South in order to exist anywhere in the South. Can you ask for a more inviting, ennobling and soul enlarging work, than that of making one of the glorious Band who shall carry Liberty to your enslaved people? Remember that identified with the Slave in color, you will have a power that white soldiers have not, to attract them to your lines and induce them to take up arms in a common cause.

One black Brigade will, for this work, be worth more than two white ones. Enlist, therefore, enlist without delay, enlist now, and forever put an end to the human barter and butchery which have stained the whole South with the warm blood of your people, and loaded its air with their groans. Enlist, and deserve not only well of your country, and win for yourselves, a name and a place among men, but secure to yourself what is infinitely more precious, the fast dropping tears of gratitude of your kith and kind marked out for destruction, and who are but now ready to perish.

When time's ample curtain shall fall upon our national tragedy, and our hillsides and valleys shall neither redden with the blood nor whiten with the bones of kinsmen and countrymen who have fallen in the sanguinary and wicked strife; when grim visaged war has smoothed his wrinkled front and our country shall have regained its normal condition as a leader of nations in the occupation and blessings of peace — and history shall record the names of heroes and martyrs — who bravely answered the call of patriotism and Liberty — against traitors, thieves and assassins — let it not be said that in the long list of glory, composed of men of all nations — there appears the name of no colored man.

Frederick Douglass (LOC)

What the Black Man Wants,
by Frederick Douglass
(1865)

Mr. President:

I came here, as I come always to the meetings in New England, as a listener, and not as a speaker; and one of the reasons why I have not been more frequently to the meetings of this society, has been because of the disposition on the part of some of my friends to call me out upon the platform, even when they knew that there was some difference of opinion and of feeling between those who rightfully belong to this platform and myself; and for fear of being misconstrued, as desiring to interrupt or disturb the proceedings of these meetings, I have usually kept away, and have thus been deprived of that educating influence, which I am always free to confess is of the highest order, descending from this platform.

I have felt, since I have lived out West, that in going there I parted from a great deal that was valuable; and I feel, every time I come to these meetings, that I have lost a great deal by making my home west of Boston, west of Massachusetts; for, if anywhere in the country there is to be found the highest sense of justice, or the truest demands for my race, I look for it in the East, I look for it here. The ablest discussions of the whole question of our rights occur here, and to be deprived of the privilege of listening to those discussions is a great deprivation.

I do not know, from what has been said, that there is any difference of opinion as to the duty of abolitionists, at the present moment. How can we get up any difference at this point, or any point, where we are so united, so agreed? I went especially, however, with that word of Mr. Phillips, which is the criticism of Gen. Banks and Gen. Banks' policy. I hold that that policy is our chief danger at the present moment; that it practically enslaves the Negro, and makes the Proclamation of 1863 a mockery and delusion. What is freedom? It is the right to choose one's own employment.

Certainly it means that, if it means anything; and when any individual or combination of individuals undertakes to decide for any man when he shall work, where he shall work, at what he shall work, and for what he shall work, he or they practically reduce him to slavery. [Applause.] He is a slave. That I understand Gen. Banks to do—to determine for the so-called freedman, when, and where, and at what, and for how much he shall work, when he shall be punished, and by whom punished. It is absolute slavery. It defeats the beneficent intention of the Government, if it has beneficent intentions, in regards to the freedom of our people.

I have had but one idea for the last three years to present to the American people, and the phraseology in which I clothe it is the old abolition phraseology. I am for the "immediate, unconditional, and universal" enfranchisement of the black man, in every State in the Union. [Loud applause.] Without this, his liberty is a mockery; without this, you might as well almost retain the old name of slavery for his condition; for in fact, if he is not the slave of the individual master, he is the slave of society, and holds his liberty as a privilege, not as a right. He is at the mercy of the mob, and has no means of protecting himself.

It may be objected, however, that this pressing of the Negro's right to suffrage is premature. Let us have slavery abolished, it may be said, let us have labor organized, and then, in the natural course of events, the right of suffrage will be extended to the Negro. I do not agree with this. The constitution of the human mind is such, that if it once disregards the conviction forced upon it by a revelation of truth, it requires the exercise of a higher power to produce the same conviction afterwards. The American people are now in tears. The Shenandoah has run blood—the best blood of the North. All around Richmond, the blood of New England and of the North has been shed—of your sons, your brothers and your fathers.

We all feel, in the existence of this Rebellion, that judgments terrible, wide-spread, far-reaching, overwhelming, are abroad in the land; and we feel, in view of these judgments, just now, a disposition to learn righteousness. This is the hour. Our streets are in mourning, tears are falling at every fireside, and under the chastisement of this Rebellion we have almost come up to the point of conceding this great, this all-important right of suffrage. I fear that if we fail to do it now, if abolitionists fail to press it now, we may not see, for centuries to come, the same disposition that exists at this moment. Hence, I say, now is the time to press this right.

It may be asked, "Why do you want it? Some men have got along very well without it. Women have not this right." Shall we justify one wrong by another? This is a sufficient answer. Shall we at this moment justify the deprivation of the Negro of the right to vote, because some one else is deprived of that privilege? I hold that women, as well as men, have the right to vote [applause.], and my heart and my voice go with the movement to extend suffrage to woman; but that question rests upon another basis than that on which our right rests. We may be asked, I say, why we want it. I will tell you why we want it. We want it because it is our *right*, first of all. No class of men can, without insulting their own nature, be content with any deprivation of their rights. We want it again, as a means for educating our race.

Men are so constituted that they derive their conviction of their own possibilities largely from the estimate formed of them by others. If nothing is expected of a people, that people will find it difficult to contradict that expectation. By depriving us of suffrage, you affirm our incapacity to form an intelligent judgment respecting public men and public measures; you declare before the world that we are unfit to exercise the elective franchise, and by this means lead us to undervalue ourselves, to put a low estimate upon ourselves, and to feel that we have no possibilities like other men.

Again, I want the elective franchise, for one, as a colored man, because ours is a peculiar government, based upon a peculiar idea, and that idea is universal suffrage. If I were in a monarchial government, or an autocratic or aristocratic government, where the few bore rule and the many were subject, there would be no special stigma resting upon me, because I did not exercise the elective franchise. It would do me no great violence.

Mingling with the mass I should partake of the strength of the mass; I should be supported by the mass, and I should have the same incentives to endeavor with the mass of my fellow-men; it would be no particular burden, no particular deprivation; but here where universal suffrage is the rule, where that is the fundamental idea of the Government, to rule us out is to make us an exception, to brand us with the stigma of inferiority, and to invite to our heads the missiles of those about us; therefore, I want the franchise for the black man.

There are, however, other reasons, not derived from any consideration merely of our rights, but arising out of the conditions of the South, and of the country—considerations which have already been referred to by Mr. Phillips—considerations which must arrest the attention of statesmen. I believe that when the tall heads of this Rebellion shall have been swept down, as they will be swept down, when the Davises and Toombses and Stephenses, and others who are leading this Rebellion shall have been blotted out, there will be this rank undergrowth of treason, to which reference has been made, growing up there, and interfering with, and thwarting the quiet operation of the Federal Government in those states. You will see those traitors, handing down, from sire to son, the same malignant spirit which they have manifested, and which they are now exhibiting, with malicious hearts, broad blades, and bloody hands in the field, against our sons and brothers. That spirit will still remain; and whoever sees the Federal Government extended over those Southern States will see that Government in a strange land, and not only in a strange land, but in an enemy's land.

A post-master of the United States in the South will find himself surrounded by a hostile spirit; a collector in a Southern port will find himself surrounded by a hostile spirit; a United States marshal or United States judge will be surrounded there by a hostile element. That enmity will not die out in a year, will not die out in an age. The Federal Government will be looked upon in those States precisely as the Governments of Austria and France are looked upon in Italy at the present moment. They will endeavor to circumvent, they will endeavor to destroy, the peaceful operation of this Government. Now, where will you find the strength to counterbalance this spirit, if you do not find it in the Negroes of the South?

They are your friends, and have always been your friends. They were your friends even when the Government did not regard them as such. They comprehended the genius of this war before you did. It is a significant fact, it is a marvelous fact, it seems almost to imply a direct interposition of Providence, that this war, which began in the interest of slavery on both sides, bids fair to end in the interest of liberty on both sides. [Applause.] It was begun, I say, in the interest of slavery on both sides.

The South was fighting to take slavery out of the Union, and the North fighting to keep it in the Union; the South fighting to get it beyond the limits of the United States Constitution, and the North fighting to retain it within those limits; the South fighting for new guarantees, and the North fighting for the old guarantees;—both despising the Negro, both insulting the Negro. Yet, the Negro, apparently endowed with wisdom from on high, saw more clearly the end from the beginning than we did. When Seward said the status of no man in the country would be changed by the war, the Negro did not believe him.

When our generals sent their underlings in shoulder-straps to hunt the flying Negro back from our lines into the jaws of slavery, from which he had escaped, the Negroes thought that a mistake had been made, and that the intentions of the Government had not been rightly understood by our officers in shoulder-straps, and they continued to come into our lines, threading their way through bogs and fens, over briers and thorns, fording streams, swimming rivers, bringing us tidings as to the safe path to march, and pointing out the dangers that threatened us. They are our only friends in the South, and we should be true to them in this their trial hour, and see to it that they have the elective franchise.

I know that we are inferior to you in some things—virtually inferior. We walk about you like dwarfs among giants. Our heads are scarcely seen above the great sea of humanity. The Germans are superior to us; the Irish are superior to us; the Yankees are superior to us [Laughter]; they can do what we cannot, that is, what we have not hitherto been allowed to do. But while I make this admission, I utterly deny, that we are originally, or naturally, or practically, or in any way, or in any important sense, inferior to anybody on this globe. [Loud applause.] This charge of inferiority is an old dodge. It has been made available for oppression on many occasions.

It is only about six centuries since the blue-eyed and fair-haired Anglo Saxons were considered inferior by the haughty Normans, who once trampled upon them. If you read the history of the Norman Conquest, you will find that this proud Anglo-Saxon was once looked upon as of coarser clay than his Norman master, and might be found in the highways and byways of Old England laboring with a brass collar on his neck, and the name of his master marked upon it were down then! You are up now. I am glad you are up, and I want you to be glad to help us up also.

The story of our inferiority is an old dodge, as I have said; for wherever men oppress their fellows, wherever they enslave them, they will endeavor to find the needed apology for such enslavement and oppression in the character of the people oppressed and enslaved. When we wanted, a few years ago, a slice of Mexico, it was hinted that the Mexicans were an inferior race, that the old Castilian blood had become so weak that it would scarcely run downhill, and that Mexico needed the long, strong and beneficent arm of the Anglo-Saxon care extended over it.

We said that it was necessary to its salvation, and a part of the "manifest destiny" of this Republic, to extend our arm over that dilapidated government. So, too, when Russia wanted to take possession of a part of the Ottoman Empire, the Turks were "an inferior race." So, too, when England wants to set the heel of her power more firmly in the quivering heart of old Ireland, the Celts are an "inferior race." So, too, the Negro, when he is to be robbed of any right which is justly his, is an "inferior man." It is said that we are ignorant; I admit it. But if we know enough to be hung, we know enough to vote.

If the Negro knows enough to pay taxes to support the government, he knows enough to vote; taxation and representation should go together. If he knows enough to shoulder a musket and fight for the flag, fight for the government, he knows enough to vote. If he knows as much when he is sober as an Irishman knows when drunk, he knows enough to vote, on good American principles.

But I was saying that you needed a counterpoise in the persons of the slaves to the enmity that would exist at the South after the Rebellion is put down. I hold that the American people are bound, not only in self-defense, to extend this right to the freedmen of the South, but they are bound by their love of country, and by all their regard for the future safety of those Southern States, to do this—to do it as a measure essential to the preservation of peace there. But I will not dwell upon this. I put it to the American sense of honor. The honor of a nation is an important thing.

It is said in the Scriptures, "What doth it profit a man if he gain the whole world, and lose his own soul?" It may be said, also, What doth it profit a nation if it gain the whole world, but lose its honor? I hold that the American government has taken upon itself a solemn obligation of honor, to see that this war—let it be long or let it be short, let it cost much or let it cost little—that this war shall not cease until every freedman at the South has the right to vote. It has bound itself to it. What have you asked the black men of the South, the black men of the whole country, to do? Why, you have asked them to incur the deadly enmity of their masters, in order to befriend you and to befriend this Government.

You have asked us to call down, not only upon ourselves, but upon our children's children, the deadly hate of the entire Southern people. You have called upon us to turn our backs upon our masters, to abandon their cause and espouse yours; to turn against the South and in favor of the North; to shoot down the Confederacy and uphold the flag—the American flag. You have called upon us to expose ourselves to all the subtle machinations of their malignity for all time. And now, what do you propose to do when you come to make peace? To reward your enemies, and trample in the dust your friends? Do you intend to sacrifice the very men who have come to the rescue of your banner in the South, and incurred the lasting displeasure of their masters thereby? Do you intend to sacrifice them and reward your enemies? Do you mean to give your enemies the right to vote, and take it away from your friends? Is that wise policy? Is that honorable?

Could American honor withstand such a blow? I do not believe you will do it. I think you will see to it that we have the right to vote. There is something too mean in looking upon the Negro, when you are in trouble, as a citizen, and when you are free from trouble, as an alien. When this nation was in trouble, in its early struggles, it looked upon the Negro as a citizen. In 1776 he was a citizen. At the time of the formation of the Constitution the Negro had the right to vote in eleven States out of the old thirteen. In your trouble you have made us citizens. In 1812 Gen. Jackson addressed us as citizens—"fellow-citizens."

He wanted us to fight. We were citizens then! And now, when you come to frame a conscription bill, the Negro is a citizen again. He has been a citizen just three times in the history of this government, and it has always been in time of trouble. In time of trouble we are citizens. Shall we be citizens in war, and aliens in peace? Would that be just?

I ask my friends who are apologizing for not insisting upon this right, where can the black man look, in this country, for the assertion of his right, if he may not look to the Massachusetts Anti-Slavery Society? Where under the whole heavens can he look for sympathy, in asserting this right, if he may not look to this platform? Have you lifted us up to a certain height to see that we are men, and then are any disposed to leave us there, without seeing that we are put in possession of all our rights?

We look naturally to this platform for the assertion of all our rights, and for this one especially. I understand the anti-slavery societies of this country to be based on two principles,—first, the freedom of the blacks of this country; and, second, the elevation of them. Let me not be misunderstood here. I am not asking for sympathy at the hands of abolitionists, sympathy at the hands of any.

I think the American people are disposed often to be generous rather than just. I look over this country at the present time, and I see Educational Societies, Sanitary Commissions, Freedmen's Associations, and the like,—all very good: but in regard to the colored people there is always more that is benevolent, I perceive, than just, manifested towards us. What I ask for the Negro is not benevolence, not pity, not sympathy, but simply *justice*.

The American people have always been anxious to know what they shall do with us. Gen. Banks was distressed with solicitude as to what he should do with the Negro. Everybody has asked the question, and they learned to ask it early of the abolitionists, "What shall we do with the Negro?" I have had but one answer from the beginning. Do nothing with us! Your doing with us has already played the mischief with us. Do nothing with us! If the apples will not remain on the tree of their own strength, if they are worm eaten at the core, if they are early ripe and disposed to fall, let them fall! I am not for tying or fastening them on the tree in any way, except by nature's plan, and if they will not stay there, let them fall. And if the Negro cannot stand on his own legs, let him fall also.

All I ask is, give him a chance to stand on his own legs! Let him alone! If you see him on his way to school, let him alone, don't disturb him! If you see him going to the dinner-table at a hotel, let him go! If you see him going to the ballot-box, let him alone, don't disturb him! [Applause.] If you see him going into a work-shop, just let him alone,—your interference is doing him a positive injury. Gen. Banks' "preparation" is of a piece with this attempt to prop up the Negro.

Let him fall if he cannot stand alone! If the Negro cannot live by the line of eternal justice, so beautifully pictured to you in the illustration used by Mr. Phillips, the fault will not be yours, it will be his who made the Negro, and established that line for his government. Let him live or die by that. If you will only untie his hands, and give him a chance, I think he will live. He will work as readily for himself as the white man. A great many delusions have been swept away by this war.

One was, that the Negro would not work; he has proved his ability to work. Another was, that the Negro would not fight; that he possessed only the most sheepish attributes of humanity; was a perfect lamb, or an "Uncle Tom;" disposed to take off his coat whenever required, fold his hands, and be whipped by anybody who wanted to whip him. But the war has proved that there is a great deal of human nature in the Negro, and that "he will fight," as Mr. Quincy, our President, said, in earlier days than these, "when there is a reasonable probability of his whipping anybody."

The Future of the Colored Race,
by Frederick Douglass
(1886)

It is quite impossible, at this early date, to say with any decided emphasis what the future of the colored people will be. Speculations of that kind, thus far, have only reflected the mental bias and education of the many who have essayed to solve the problem.

We all know what the negro has been as a slave. In this relation we have his experience of two hundred and fifty years before us, and can easily know the character and qualities he has developed and exhibited during this long and severe ordeal. In his new relation to his environments, we see him only in the twilight of twenty years of semi-freedom; for he has scarcely been free long enough to outgrow the marks of the lash on his back and the fetters on his limbs. He stands before us, to-day, physically, a maimed and mutilated man. His mother was lashed to agony before the birth of her babe, and the bitter anguish of the mother is seen in the countenance of her offspring.

Slavery has twisted his limbs, shattered his feet, deformed his body and distorted his features. He remains black, but no longer comely. Sleeping on the dirt floor of the slave cabin in infancy, cold on one side and warm on the other, a forced circulation of blood on the one side and chilled and retarded circulation on the other, it has come to pass that he has not the vertical bearing of a perfect man. His lack of symmetry, caused by no fault of his own, creates a resistance to his progress which cannot well be overestimated, and should be taken into account, when measuring his speed in the new race of life upon which he has now entered.

As I have often said before, we should not measure the negro from the heights which the white race has attained, but from the depths from which he has come. You will not find Burke, Grattan, Curran and O'Connell among the oppressed and famished poor of the famine-stricken districts of Ireland. Such men come of comfortable antecedents and sound parents

Laying aside all prejudice in favor of or against race, looking at the negro as politically and socially related to the American people generally, and measuring the forces arrayed against him, I do not see how he can survive and flourish in this country as a distinct and separate race, nor do I see how he can be removed from the country either by annihilation or expatriation.

Sometimes I have feared that, in some wild paroxysm of rage, the white race, forgetful of the claims of humanity and the precepts of the Christian religion, will proceed to slaughter the negro in wholesale, as some of that race have attempted to slaughter Chinamen, and as it has been done in detail in some districts of the Southern States. The grounds of this fear, however, have in some measure decreased, since the negro has largely disappeared from the arena of Southern politics, and has betaken himself to industrial pursuits and the acquisition of wealth and education, though even here, if over-prosperous, he is likely to excite a dangerous antagonism; for the white people do not easily tolerate the presence among them of a race more prosperous than themselves.

The negro as a poor ignorant creature does not contradict the race pride of the white race. He is more a source of amusement to that race than an object of resentment. Malignant resistance is augmented as he approaches the plane occupied by the white race, and yet I think that that resistance will gradually yield to the pressure of wealth, education, and high character.

My strongest conviction as to the future of the negro therefore is, that he will not be expatriated nor annihilated, nor will he forever remain a separate and distinct race from the people around him, but that he will be absorbed, assimilated, and will only appear finally, as the Phoenicians now appear on the shores of the Shannon, in the features of a blended race. I cannot give at length my reasons for this conclusion, and perhaps the reader may think that the wish is father to the thought, and may in his wrath denounce my conclusion as utterly impossible. To such I would say, tarry a little, and look at the facts.

Two hundred years ago there were two distinct and separate streams of human life running through this country. They stood at opposite extremes of ethnological classification: all black on the one side, all white on the other. Now, between these two extremes, an intermediate race has arisen, which is neither white nor black, neither Caucasian nor Ethiopian, and this intermediate race is constantly increasing. I know it is said that marital alliance between these races is unnatural, abhorrent and impossible; but exclamations of this kind only shake the air.

They prove nothing against a stubborn fact like that which confronts us daily and which is open to the observation of all. If this blending of the two races were impossible we should not have at least one-fourth of our colored population composed of persons of mixed blood, ranging all the way from a dark-brown color to the point where there is no visible admixture. Besides, it is obvious to common sense that there is no need of the passage of laws, or the adoption of other devices, to prevent what is in itself impossible.

Of course this result will not be reached by any hurried or forced processes. It will not arise out of any theory of the wisdom of such blending of the two races. If it comes at all, it will come without shock or noise or violence of any kind, and only in the fullness of time, and it will be so adjusted to surrounding conditions as hardly to be observed. I would not be understood as advocating intermarriage between the two races.

I am not a propagandist, but a prophet. I do not say that what I say *should* come to pass, but what I think is likely to come to pass, and what is inevitable. While I would not be understood as advocating the desirability of such a result, I would not be understood as deprecating it. Races and varieties of the human family appear and disappear, but humanity remains and will remain forever. The American people will one day be truer to this idea than now, and will say with Scotia's inspired son:

"A man's a man for a' that."

When that day shall come, they will not pervert and sin against the verity of language as they now do by calling a man of mixed blood, a negro; they will tell the truth. It is only prejudice against the negro which calls every one, however nearly connected with the white race, and however remotely connected with the negro race, a negro. The motive is not a desire to elevate the negro, but to humiliate and degrade those of mixed blood; not a desire to bring the negro up, but to cast the mulatto and the quadroon down by forcing him below an arbitrary and hated color line.

Men of mixed blood in this country apply the name "*negro*" to themselves, not because it is a correct ethnological description, but to seem especially devoted to the black side of their parentage. Hence in some cases they are more noisily opposed to the conclusion to which I have come, than either the white or the honestly black race. The opposition to amalgamation, of which we hear so much on the part of colored people, is for most part the merest affectation, and, will never form an impassable barrier to the union of the two varieties.

Ain't I a Woman?,
by Sojourner Truth
(1851)

This speech was given to the Women's Rights Convention in Akron, Ohio in May of 1851. As recorded by Marcus Robinson, newspaper editor.

I want to say a few words about this matter. I am a woman's rights. I have as much muscle as any man, and can do as much work as any man. I have plowed and reaped and husked and chopped and mowed, and can any man do more than that? I have heard much about the sexes being equal. I can carry as much as any man, and can eat as much too, if I can get it. I am as strong as any man that is now. As for intellect, all I can say is, if a woman have [sic] a pint, and a man a quart — why can't she have her little pint full? You need not be afraid to give us our rights for fear we will take too much, — for we can't take more than our pint'll [sic] hold. The poor men seems [sic] to be all in confusion, and don't know what to do.

Why children, if you have woman's rights, give it to her and you will feel better. You will have your own rights, and they won't be so much trouble. I can't read, but I can hear. I have heard the bible and have learned that Eve caused man to sin. Well, if woman upset the world, do give her a chance to set it right side up again.

The Lady has spoken about Jesus, how he never spurned woman from him, and she was right. When Lazarus died, Mary and Martha came to him with faith and love and besought him to raise their brother. And Jesus wept and Lazarus came forth. And how came Jesus into the world? Through God who created him and the woman who bore him. Man, where was your part? But the women are coming up blessed be God and a few of the men are coming up with them. But man [sic] is in a tight place, the poor slave is on him, woman is coming on him, he is surely between a hawk and a buzzard.

Sojourner Truth (LOC)

The Interview with Mr. Lincoln,
by Sojourner Truth
(1864)

Mrs. Truth met with President Abraham Lincoln on October 29, 1864. A copy of her rendition of the meeting was published in the Book of Life *section of the* Narrative in 1875.

It was about 8 o'clock A.M., when I called on the president. Upon entering his reception room we found about a dozen persons in waiting, among them two colored women. I had quite a pleasant time waiting until he was disengaged, and enjoying his conversation with others; he showed as much kindness and consideration to the colored persons as to the whites -- if there was any difference, more.

One case was that of a colored woman who was sick and likely to be turned out of her house on account of [sic] her inability to pay her rent. The president listened to with much attention, and spoke to her with kindness and tenderness. He said he had given so much he could give no more, but told her where to go and get the money, and asked Mrs. C---n to assist her, which she did.

The president was seated at his desk. Mrs. C. said to him, "This is Sojourner Truth, who has come all the way from Michigan to see you." He then arose, gave me his hand, made a bow, and said, "I am pleased to see you."

I said to him, Mr. President, when you first took your seat I feared you would be torn to pieces, for I likened you unto Daniel, who was thrown into the lion's den; and if the lions did not tear you into pieces, I knew that it would be God that had saved you; and I said, if he spared me I would see you before the four years expired, and he has done so, and now I am here to see you for myself.

He then congratulated me upon having been spared. Then I said, I appreciate you, for you are the best president who has ever taken the seat. He replied: 'I expect you have reference to my having emancipated the slaves in my proclamation. But,' said he, mentioning the names of several of his predecessors (and among them emphatically that of Washington), 'they were all just as good, and would have done just as I have done if the time had come.

If the people over the river [pointing across the Potomac] had behaved themselves, I could not have done what I have; but they did not, which gave the opportunity to do those things.' I then said, I thank God that you were the instrument selected by him and the people to do it. I told him that I had never heard of him before he was talked of for president. He smilingly replied, 'I had heard of you many times before that.'

He then showed me the Bible presented to him by the colored people of Baltimore, of which you have no doubt seen a description. I have seen it for myself and it is beautiful beyond description.

After I had looked it over, I said to him, This is beautiful indeed; the colored people have given this to the head of the government, and that government once sanctioned laws that would not permit its people to learn enough to enable them to read this book. And for what? Let them answer who can.

I must say, and I am proud to say, that I never was treated by any one with more kindness and cordiality than were shown to me by that great and good man, Abraham Lincoln, by the grace of God president of the United States for four years more. He took my little book, and with the same hand that signed the death-warrant of slavery, wrote as follows:

As I was taking my leave, he arose and took my hand, and said he would be pleased to have me call again. I felt that I was in the presence of a friend, and now I thank God from the bottom of my heart that I always have advocated his cause, and have done it openly and boldly. I shall feel still more in duty bound to do so in time to come. May God assist me.

President Lincoln showing Sojourner Truth his Bible (LOC)

On the Injustice of Slavery
by Sojourner Truth
(1856)

This speech was given to the "Friends of Human Progress Association." Shortly after this speech, Mrs. Truth moved to Michigan, where she lived until her death in 1883. The acting secretary of the Association, Thomas Chandler recorded this speech and it was also published in the Anti-Slavery *periodical, The Bugle in October of 1856.*

   As you were speaking this morning of little children, I was looking around and thinking it was most beautiful. But I have had children and yet never owned one, no one ever owned one; and of such there's millions -- who goes to teach them? You have teachers for your children but who will teach the poor slave children?

I want to know what has become of the love I ought to have for my children? I did have love for them, but what has become of it? I cannot tell you. I have had two husbands but I never possessed one of my own. I have had five children and never could take one of them up and say, 'My child' or 'My children,' unless it was when no one could see me.

I believe in Jesus, and I was forty years a slave but I did not know how dear to me was my posterity. I was so beclouded and crushed. But how good and wise is God, for if the slaves knowed [sic] what their true condition was, it would be more than the mind could bear. While the race is sold of all their rights -- what is there on God's footstool to bring them up? Has not God given to all his creatures the same rights? How could I travel and live and speak? When I had not got something to bear me up, when I've been robbed of all my affections for husband and children.

Some years ago there appeared to me a form (here the speaker gave a very graphic description of the vision she had). Then I learned that I was a human being. We had been taught that we was [was] a species of monkey, baboon or 'rang-o-tang, and we believed it -- we'd never seen any of these animals. But I believe in the next world. When we gets [sic] up yonder, we shall have all of them rights 'stored to us again -- all that love what I've lost -- all going to be 'stored to me again. Oh! How good God is.

My mother said when we were sold, we must ask God to make our masters good, and I asked who He was. She told me, He sit up in the sky. When I was sold, I had a severe, hard master, and I was tied up in the barn and whipped. Oh! Till the blood run down the floor and I asked God, why don't you come and relieve me -- if I was you and you'se [sic] tied up so, I'd do it for you.

Battle Hymn of the Republic,
by Julia Ward Howe
(1861)

Mine eyes have seen the glory of the coming of the Lord
He is trampling out the vintage where the grapes of wrath are stored,
He has loosed the fateful lightening of His terrible swift sword
His truth is marching on.

Glory! Glory! Hallelujah!

Glory! Glory! Hallelujah!

Glory! Glory! Hallelujah!

His truth is marching on.

I have seen Him in the watch-fires of a hundred circling camps
They have builded Him an altar in the evening dews and damps
I can read His righteous sentence by the dim and flaring lamps
His day is marching on.

Cover of the 1862 sheet music for "The Battle Hymn of the Republic"

Glory! Glory! Hallelujah!

Glory! Glory! Hallelujah!

Glory! Glory! Hallelujah!

His truth is marching on.

I have read a fiery gospel writ in burnish'd rows of steel,
"As ye deal with my contemners, So with you my grace shall deal;"

Let the Hero, born of woman, crush the serpent with his heel
Since God is marching on.

Glory! Glory! Hallelujah!

Glory! Glory! Hallelujah!

Glory! Glory! Hallelujah!

His truth is marching on.

He has sounded forth the trumpet that shall never call retreat.
He is sifting out the hearts of men before His judgment-seat.
Oh, be swift, my soul, to answer Him! be jubilant, my feet!
Our God is marching on.

Glory! Glory! Hallelujah!

Glory! Glory! Hallelujah!

Glory! Glory! Hallelujah!

His truth is marching on.

In the beauty of the lilies Christ was born across the sea,
With a glory in His bosom that transfigures you and me:
As He died to make men holy, let us die to make men free,
While God is marching on.

Glory! Glory! Hallelujah!

Glory! Glory! Hallelujah!

Glory! Glory! Hallelujah!

His truth is marching on.

Chapter 12

The Regiments

During the Civil War, both the rank structure and the organizational unit structure was somewhat different than today. Additionally, regarding African American soldiers, the rules of their service were different from their white counterparts. These "Colored" units were commanded by white officers but some blacks served as junior grade officers (Lieutenants and Captains). These regiments were called United States Colored Troops (U.S.C.T.). Even in the North, racial inequality remained contentious social and political issues. A review of the Union Army is provided as it did have a bearing on African Americans.

The Highest-ranking commander of the Union Army was designated as the "General-in-Chief" during the Civil War. These senior officers were Lieutenant Generals (three stars) and they were appointed by the president and confirmed by the Senate. President Abraham Lincoln had four during his administration (1861-1865): **Winfield Scott: 1861; George B. McClellan: 1861-1862; Henry W. Halleck: July 23, 1862-1864; Ulysses S. Grant: March 9, 1864- 1865** and during Reconstruction till 1869.

22nd Regimental Colors (LOC)

When the war began in 1861, some Northern states already had state militia units and these were incorporated into the U.S. Army and designated as federal regiments. Prior to 1863, several thousand African American troops served as support troops but they were not allowed to fight. In 1863, the War Department issued General Order 143, which authorized the United States Army to raise African American regiments to serve as U.S.C.T units alongside white regiments. The U.S.C.T. regiments were assigned to various units as directed by the U.S. Army.

During the Civil War era, the regimental system was the basic unit structure for senior officers to deploy and maneuver units to and from various locations and duties depending on their assigned missions. The regiments filled multiple "Armies," Corps, Divisions and Brigades, which were arranged within various geographic departments during the war.

Roughly 186,000 African Americans served in 175 U.S.C.T. regiments as volunteers in Infantry, Cavalry, and Artillery units. Again, there were thousands of other support troops assigned to various support units (medical, quartermaster supply, teamsters, etc.) but these soldiers performed common labor duties and they did not fight as organized forces.

During the Civil War, except for the "General-in-Chief," the Union Army only had two grades of general: Major General (two stars) and Brigadier General (one star). The basic field organizational composition of the U.S. Army was structured (from the top-down) as follows:

Army: comprised of two or more Corps and commanded by a Major General. The U.S. Army during the Civil War was comprised of ten Armies. These Armies were: Army of the Cumberland; Army of Georgia; Army of the Gulf; Army of the James; Army of the Mississippi; Army of the Ohio; Army of the Potomac; Army of the Shenandoah; Army of the Tennessee; and the Army of Virginia.

Corps: comprised of two or more Divisions (roughly 8,000 men). Commanded by a Major General or a Brigadier General.
Division: comprised of two or more Brigades (roughly 4,000 men). Commanded by a Major General, but some were commanded by Brigadier Generals.
Brigade: two or more Regiments (2,000 men or more). Commanded by a Brigadier General or a senior Colonel.
Regiment: Comprised of roughly 1,000 men.
Company: Comprised of roughly 100 men.

Each regiment had a command and staff structure comprised of: 1 Colonel, 1 Lieutenant Colonel, 1 Major, 1 Adjutant, 1 Quartermaster, 1 Surgeon, 2 Assistant Surgeons, 1 Chaplain, 1 Sergeant Major, 1 Quartermaster Sergeant, 1 Commissary Sergeant, 1 Hospital Steward, and 1 Principal Musician.

Each company contained roughly 100 men. Leadership in the companies were comprised primarily of white officers, in the following rank structure:

1 Captain, 1 First Lieutenant, 1 Second Lieutenant, 1 First Sergeant, 4 Sergeants, 8 Corporals, 2 Musicians, 1 Wagoner, and 64-80 Privates.

26th U.S. Colored Regiment (NYS)

The following regiments were all U.S. Colored troops and assigned to various corps, divisions and brigades as needed throughout the U.S. Army. During the war, units were sometimes reorganized due to the needs of the U.S. Army (casualties, different unit formations, different missions, etc.), as changes occurred.

Infantry

135 regiments of Infantry (1st-138th US Colored Infantry)

Notes:
1 unassigned Company of Infantry (Company A, US Colored Infantry)

1 Independent USC Company of Infantry (Southard's Independent (Colored) Company, Pennsylvania (Colored) Infantry
Independent US Colored Regiment of Infantry (Powell's Regiment, US Colored Infantry)
The 94th, 105th, and 126th US Colored Infantry regiments never achieved fully manned status

Notes: Reorganized Infantry Units
5th Regiment Infantry, U.S. Colored Troops- Originally the 127th Ohio Volunteer Infantry Regiment
33rd Regiment Infantry, U.S. Colored Troops- Originally the 1st South Carolina Colored Volunteers
54th Regiment Infantry, U.S. Colored Troops - Originally, the Massachusetts Volunteer Infantry
62nd Regiment Infantry, U.S. Colored Troops - Originally, 1st Missouri Regt of Colored Infantry
65th Regiment Infantry, U.S. Colored Troops - Originally, 2nd Missouri Regt of Colored Infantry
67th Regiment Infantry, U.S. Colored Troops - Originally, 3rd Missouri Regt of Colored Infantry
68th Regiment Infantry, U.S. Colored Troops - Originally, 4th Missouri Regt of Colored Infantry
73rd Regiment Infantry, U.S. Colored Troops - Originally, 1st- 2nd Corps d' Afrique Infantry
76th Regiment Infantry, U.S. Colored Troops- Originally, 3rd- 4th Corps d' Afrique Infantry
On April 4, 1864, the 1st–20th, 22nd, and 26th Corps d' Afrique Infantry were reorganized as the 77th-79th, 80th-83rd, 84th-88th, and 89th-93rd US Colored Infantry.

Cavalry

6 Regiments (Troops) of Cavalry (1st–6th US Colored Cavalry)

Notes: Reorganized Cavalry Units
In 1864, the 1st Corps d' Afrique Cavalry was reorganized as the 4th US Colored Cavalry

Artillery

13 Heavy Artillery Regiments (1st and 3rd–14th US Colored (Heavy) Artillery)
1 Regiment of Light Artillery (2nd US Colored (Light) Artillery)
1 Independent US Colored (Heavy) Artillery Battery

Notes: Reorganized Artillery Units
2nd Regiment Heavy Artillery African Descent - Reorganized as the 4th Regiment U.S. Colored.
8th Regiment Heavy Artillery U.S. Colored Troops - Originally, 1st Alabama Siege Artillery Colored.
11th Regiment Heavy Artillery U.S. Colored Troops- Originally, 14th Rhode Island Heavy Artillery Colored
1 Regiment of Louisiana Heavy Artillery- Reorganized as the 10th US Colored (Heavy) Artillery

26th Regimental Colors. *The motto reads "God and Liberty." The silk national flag includes 35 embroidered stars and an embroidered designation, "26th Regt. U.S. Colored Troops," along its center red stripe. (Courtesy of New York State Military Museum)*

Bibliography

Atkinson, Edward, ed. *The Monument to Robert Gould Shaw: Its Inception, Completion and Unveiling 1865-1897*. Boston, 1897.

Blight, David W. *Frederick Douglass' Civil War: Keeping Faith in Jubilee.* Baton Rouge: Louisiana State University Press, 1989.

Blight, David. "The Meaning or the Fight: Frederick Douglass and the Memory of the Fifty-fourth Massachusetts." *The Massachusetts Review* 36 (Spring 1995): 141-153.

Brown, Hallie Q. (Hallie Quinn). *Homespun Heroines and Other Women of Distinction*. Xenia, Ohio: Aldine Publishing Co., 1926.

Burchard, Peter, *One Gallant Rush: Robert Gould Shaw and His Brave Black Regiment*. New York: St. Martin's Press, 1989.

Burkhardt, J. Peter. *All Made of Tunes: Charles Ives and the Uses of Musical Borrowing*. New Haven, 1995.

Burton, Annie L. *Memories of Childhood's Slavery Days*. Boston: Ross Publishing Company, 1909.

Child, Lydia Maria Francis, Ed. *The Freedmen's Book*. Boston: Ticknor and Fields, 1865.

Coffin, William; and Thomas Wentworth Higginson. "The Shaw Memorial and the Sculptor St. Gaudens." *Century Magazine* 54 (June 1897): 176-200.

Cowley, Charles, *The Romance of History in "the Black County," and the Romance of War in the Career of Gen. Robert Smalls*. Lowell, Mass: 1882.

Dameron, J. David, *Horace King: From Salve to Master Builder and Legislator.* Southeast Research Publishing, LLC. 2017.

Dryfhout, John H. *The Work of Augustus Saint-Gaudens*. Hanover, N.H. and London, 1982.

Donald, David *et al. The Civil War and Reconstruction*. 2001.

Douglas, Frederick, *Life and Times of Frederick Douglass*. Boston: De Wolfe & Fiske Co., 1892.

Douglass, Frederick. *Escape From Slavery.* Edited by Michael McCurdy. New York: Knopf, 1994.

Douglass, Frederick. *My Bondage and My Freedom.* New York: Miller, Orton and Mulligan, 1855. Reprint, Urbana: University of Illinois Press, 1987.

Douglass, Frederick. *Narrative of the Life of Frederick Douglass.* Boston: Anti-slavery Office, 1845. Reprint, New Brunswick, NJ: Transaction Publishers, 1997.

Duncan, Russell, Ed. *Blue-Eyed Child of Fortune: The Civil War Letters of Colonel Robert Gould Shaw*. Athens, GA: University of Georgia Press, 1992.

Eicher, David J. *The Longest Night: A Military History of the Civil War.* New York: Simon & Schuster, 2001.

Emilio, Luis F. *History of the Fifty-Fourth Regiment of Massachusetts Volunteer Infantry, 1863-1865.* New York, 1968.

Fellman, Michael *et al. This Terrible War: The Civil War and its Aftermath* (2nd. ed. 2007).

Gooding, James Henry. Edited by Virginia Adams. *On the Altar of Freedom: A Black Soldier's Civil War Letters from the Front*. New York, 1992.

Grigsby, Darcy Grimaldo. *Enduring Truths: Sojourner's Shadows and Substance.* Chicago: University of Chicago Press, 2015.

Guelzo, Allen C. *Fateful Lightning: A New History of the Civil War & Reconstruction.* New York: Oxford University Press, 2012.

Hill, Lawrence. *Trials and triumphs: the story of African-Canadians.* Lawrence Hill - Toronto: Umbrella Press, 1993.

Jacobs, Harriet A. *Incidents in the Life of a Slave Girl.* Boston: The Author, 1861.

Kammen, Michael G., *Digging Up the Dead: A History of Notable American Reburials*. Chicago: University of Chicago Press, 2010.

Keckley, Elizabeth. *Behind the Scenes: Thirty Years a Slave, and Four Years in the White House.* New York: *G. W. Carleton & Co., Publishers. 1868.*

Larson, Kate Clifford. *Bound for the Promised Land: Harriet Tubman, Portrait of an American Hero*, Ballantine. New York:

McPherson, James M. *Battle Cry of Freedom: The Civil War Era*. New York: Oxford University Press, 1988.

Sprague, Rosetta Douglass, *Anna Murray Douglass, My Mother As I Recall Her.* 1900. Rpt. Fredericka Douglass Sprague Perry, 1923.

Sterling, Dorothy. *The Making of an Afro-American: Martin Robison Delany 1812–1885*, 1971, reprint Da Capo Press, 1996.

Taylor, Susie King. *Reminiscences of My Life in Camp with the 33D United States Colored Troops Late 1st S.C. Volunteers.* Boston: 1902.

List of Illustrations

Abbreviation Key:

GC: *Greene Collection*

LOC: *Library of Congress, Prints and Photograph Division*

LOCMD: *Library of Congress, Map Division*

NARA: *National Archives and Records Administration*

NHC: *U.S. Naval History Center*

NPG: *National Portrait Gallery*

Front Matter:

iii- LOC, LC-DIG-ppmsca-26454; iv-LOC, 26463; v-LOC, 27027; v, LOC-10860; vii-LOC, 90345; viii, LOC, 44265

Chapter 1:

1-NPG, 2002-89; 2,LOC, 9850440; 3, M, LOC 11398; 4, T, NPG, 79.229; 5, T, LOC, U5Zc4-4543; 5, B, GC-101; 6, R-LOC, 96522312; 7, LOC-TR,57019; 7, TL, LOC, 03351; 7, BL, 57020; 7, LOC, BR, 00057; 9, T, Courtesy of Columbus Museum; 9, B, GC-102; 10, L, GC-103; 10, R, GC-104; 11, C, Courtesy of Brooklyn Museum; 11, BL, LOC, 11360; 11, BR, GC-105; 12, R, GC-106; 13, T, GC-121; 13, M-United States Census Bureau (USCB)-1.

Chapter 2:

16, M, LOC 11338; 17, T, LOC, 34829; 18, R-Courtesy University of Toronto-1; 19, M, Courtesy National Library of Medicine-1; 20, R, Courtesy University of Toronto-2; 21, C, LOC, 01005; 22, TL, GC-105; 22, LOC, 01022; 23, TL, LOC, 11196; 23, TR, LOC 11182; 24, R, LOC, 19235; 25, T, GC-106; 26, BL, LOC, 03979; 26, BR, LOC, 54852; 27, T, GC-107; 27, B, GC-107.

Chapter 3:

30, TL, NPG, 74.75; 30, TR, LOC, 83188; 31, R, GC-111; 32, M, GC-112; 33, M, NARA, 5-987-44; 34, L, NARA, 8-665-21; 34, R, NARA, 8-546-76; 35, R, NARA, 6-329-65; 36, M, Harper's Weekly; 37, M, LOC 03372; 38, M, LOC, 64872; 39, T, LOC, 27188; 39, B, GC-114; 40, M, LOC, 00218; 41, M, USCB-2.

Chapter 4:

40, BM, LOC, 10980; 44, M, LOC, 11269; 45, BR, NPG, 76-101; 46, M, LOC, 02781; 47, C, LOC, B811-2553[P&P]; 48, C, Courtesy of US Army.

Chapter 5:

49, M, LOC, DIG-pga-04035; 51, TL, LOC, 3e18561; 51, TR, LOC, 3c18556; 51, BL, LOC, GC-114; 51, BR, LOC, 63886; 52, M, LOC, DIG, cwpb-01947; 54, M, Harper's Weekly; 55, M, NHC, 55510; 56, T, NARA, NWDNS-111-B-2011; 56, B, LOC, 11292; 57, M, NHC, 73688; 58, T, LOC, 01983; 58, B, LOC, 10886; 59, M, LOC, 04294; 60, T, GC-118; 60, B, GC-133.

Chapter 6:

61 B, NARA, 5c-8854; 63, Harper's Weekly; 64, TL, LOC, 85592; 64, TR, LOC, 11300; 64, BL, LOC, 13484; 64, BR, LOC, 32088; 65, TL, LOC, 26880; 65, TR, LOC, 34363; 65, BR, LOC, 50221; 65, BL, LOC, 34355; 66, TL, LOC, 32652; 66, TR, LOC, 27091; 66, BL, LOC, 32668; 66, BR, LOC, 71552; 67, BL, LOC, 11524; 67, TL, LOC, 50169; 67, BL, LOC, 26988; 67, TR, LOC, 11310; 68, TL, LOC, 11280; 68, TR, LOC, 26959; 68, BR, LOC, 34366; 68, BL, LOC, 11520; 69, BR, LOC, 27520; 69, TL, LOC, 27027; 69, TR, LOC, 26456; 69, BL, LOC, 27532; 70, BR, LOC, 27295; 70, TR, LOC, 32107; 70, TL, LOC, 27014; 70, TL, LOC, 11004; 70, BR, LOC, 11042; 71, BL, LOC, 11038; 71, TR, LOC, 10868; 71, TL, LOC, 14022; 71, BR, LOC,14022; 72, BL, LOC, 11184; 72, BR, LOC, 842966; 72, TL, LOC, 00821; 72, TR, LOC, 01062; 73, T, LOC, 82276; 73, B, LOC, 66933; 74, TC, LOC, 72256.

Chapter 7:

75, C, LOC, 3c4256; 76, B, LOC, 10896; 80, C, LOC, 02020; 80, C, LOC, 73362; 81, M, LOC, 00230; 82, Jarek Tuszyński / CC-BY-SA-3.0 & GDFL; 83, M, NARA, 67-743-21.

Chapter 8:

86, C, LOC, 11194; 87, M, LOC, 134622; 87, C, LOC, 62775; 89, C, LOC, 02043; 90, C, LOC, 39544; 91, M, LOC, 3c19848; 92, From Reminiscences of My Life in Camp; 94, C, Frank Leslie's Illustrated, 1863; 96, C, Harper's Weekly; 97, M, LOC, 04324.

Chapter 9:

99, BR, LOC, 3a10453; 100, C, Courtesy Cincinnati Art Museum; 101, C, LOC, 00468; 102, C, Harper's Weekly; 103, C, LOC, 3g02642v; 104, TC, Courtesy of New York Times; 105, BR, State of Mississippi; 107, C, LOC, 00554; 109, C, LOC, 132913; 110 C, LOC, 30088.

Chapter 10:

113, B, Harper's Weekly.

Chapter 11:

149, C, LOC, 62962; 153, T, LOC 72265; 162, C, 77293; 164, C, LOC, 55826; 166, T, LOC, 67228

Chapter 12:

169, C, LOC, 3a24164; 171, C, The image of the 26th USCT in formation, from the collection of New York State Military Museum and Veterans Research Center in Saratoga Springs, comes from www.correctionhistory.org whose page about that unit includes extensive details about its history. Courtesy of Tom McCarthy; 173, C, CM. 2005.0152 and CM. 2000.0076, New York State Military Museum and Veterans Research Center.

Index

A

Abbot, Anderson R.	iii, v,18,22
Abolitionist	1-12, 28-36,41-42,59,94,113,145-149
African Methodist Episcopal (AME) Church	5, 99-100
African Methodist Zion Church	28, 99
Alabama, Montgomery	i, 104
Alabama, Girard (Phenix City)	8
Alabama State House	10
Albumen print	vi
Ambrotype	v-vi
Amputate (amputee)	18,50,80,139
Anderson, Aaron	47
Anderson, Bruce	51
Angel of Mercy	81,84
Anthony, Susan B.	97,99
Arms (armed)	30-31,91,115-116,126,138,142,144
assassination	19,23-24,106,144

B

Ballot	32,91,101,149
Baltimore, Maryland	6,20,27,100,116,127,155
Battle Hymn of the Republic	159
Barker, Camp	15-16

Barnes, William	50
Barton, Clara	84
Baumfree, Isabella (see Sojourner Truth)	iii, 1-28,154-157
Beatty, Powhatan	48,50
Bias	130,151
Blake, Robert	47
Bone saw	18
Brady, Matthew	21,102
Bridge builder	iii, 8-10
Broadside	ix,5,12,20,37
Bronson, James	50
Brown, John	12,30
Brown, William H.	47
Brown, Wilson	47
Bull Run, 1st Battle of, Virginia	30,104
Business	8-9,20,24,30,41,111

C

Cabinet card	vi,59
Canada	v,16-18,30
Carney, William H.	iii,48,51,72-73,75,105
Camp	15-16,35,40,49,60,81-89,95
Carter, Josiah	54
Casualty	77-78,85,189
Cemetery	17,78,91,99,101
Census	12-13,39
Chaffins Farm, Battle of	48-51,105
Chicago World's Fair	25,127
Citizen (citizenship)	iv,2-3,12,17,30,32,46,58,87,99,102

Civil War	12-17,24-34,42,58,77,81,102
Color bearer	50-53,75
Colored (people)	6,16,19,25-31,42-60,80-82,89,97
Colored Troops, United States	6,16,19,25-31,42-60,80-82,89,97
Combahee river raid	97
Confederacy	14,30,33-36,76,115,137,148
Confederate	9,13-14,33-37,49-51,74-78,86,101,104
Confidante	21
Congress	1-2,14,17,36,60,72-75,99-100,109,117
Connecticut, 29th Colored Infantry Regiment	35
Conscript	9,104,149
Contraband	15-22,38,47,58,96
Constitution	30,32,90,102,113,121,129,149
Corps d' Afrique	60
Courage	29,47,72,79,104
Covey, Edward	27
C.S.S. Planter	34
C.S.S. Tennessee	47,49

D

Daguerre, Louis (Daguerreotype)	v-vi
Davis, Jefferson President	14,20,36,104
Davis, Nelson	97,99
Davis, Varina	14,20,36,104
Decorated (decoration)	46,49
Defenses	74,105
Degradation	40,42,81,124-128
Delany, Martin R.	40,42,124-128
Died (death)	10,17,23,25,42,77,81,88,91,99,101

Dorsey, Decatur	51
Douglass, Anna	27-28,33
Douglass, Charles Remond	31-32
Douglass, Frederick	21,27-3341-42,46,58,107-151
Douglass, Frederick Jr.	31
Douglass, Lewis	31-32
Draft (conscription)	104-105
Dred Scott decision	3
Dressmaker (dress making)	19-21

E

Election	32,94,101,104,106,140
Emancipation	7,40,42,90,104,119,140,144
Emancipation Proclamation	36,39-40,104,140,145,155
Equality (equal protection, equal rights)	6,17,28,30,32,42

F

Federal (government, troops)	12,14,30,82,91,106,147
First Lady	19-23
Fleetwood, Christian	48,50
Florida, Jacksonville	90
Fort Gilmer	105
Fort Harrison	105
Fort Wagner	34,48,51,55,73-78,86,90,105
Fourteenth Amendment	32,102
Fifteenth Amendment	32,102
Freedom	4,18,20-30,36,59,81,86,94,101-102
Freedman (bureau)	6,149

Fugitive Slave Law 1-2,117-118,143

G

Gardner, James 46,50
Garrison, Frank Lloyd 29-30,41,122,129
Georgia, Atlanta 105
Georgia, Columbus 8-9
Georgia, LaGrange 9-10
Georgia, Red Oak Creek 9
Georgia, Savannah 81,87,105
Gladstone, William v, 60
Godwin, John 8
"Gordon" (see Whipped Peter) 1,59-60
Governor Robert Scott 90
Gown 21,25
Grand Review 106
Great Britain 12,30,109

H

Hamburg Riot (Massacre) 91
Harper's Weekly (periodical) 34,51,59-60,92,97,106
Harris, James H. 48,50,53
Harrison, President Benjamin 33
Hawkins, James 50
Hayes, Rutherford B. President 91
Higginson, Thomas Wentworth Colonel 40,88-89,161
Holland, Milton 50
Honey Hill, Battle of 51,85,90

Hospital	6,17,19,84
Howard University	17
Howe, Julia Ward	58
Howland, Emily	97

I

Integrated	35,52,71
Ireland	30,121,128,134
Ironclad	54

J

James Island, Battle of	85,90
Johnson, Eastman	11
Jones, Miles	50

K

Kansas	3
Keckley, Elizabeth	iii,19-26,162
Keckley, George	20,23
Kelly, Alexander	50
Kentucky	i,12
King, Horace	iii-v,7-10,161
King bridge builders	7-10
Ku Klux Klan	91

L

Landsman	47
Lawson, John Henry	47

Lee, Robert E. General (CSA)	14,20,104
Legislature (legislative, legislator)	3,9,34,90,100-102
Legree, Simon	12
Library of Congress	i-vii,60
Liljenquist Family Collection	iv-vii,60
Lincoln, Abraham President	6,12-14,23,30,37,104-105,140
Lincoln, Mary Todd	21,25
Lincoln, Thomas "Tad"	23
Lincoln, William "Willie"	23
Louisiana, Baton Rouge	59,161
Louisiana, 1st Native Guards	60
Louisiana, Corps d' Afrique	60

M

Married (marriage)	8-9,27,33,41,77,82,87,91,94,97,100
Martin, Lewis	80
Maryland, Antietam	104
Maryland, Baltimore	6,20,27,100,116
Massachusetts, 5th Colored Cavalry	31
Massachusetts, 54th Colored Infantry Regiment	6,31-32,48,51,55,57,74-79,105,142
Massachusetts, 55th Colored Infantry Regiment	51
Massachusetts, Boston	59,87,119,122,127,145
Medal of Honor	45-51,72-74,104-105
Medical school	16-17,41
Medicine	17-18,41
Memorial	25,79,103,161
Mifflin, James	49
Militia	14,91,104,131
Minister	33,100-101

Mississippi, State of	100-102,105-106,140
Mississippi, Natchez	100
Mississippi, Vicksburg	100,105
Missouri, St. Louis	20,127
Mobile Bay, Battle of	46-47,49,104
Modiste	21
Montgomery, James Colonel	97
Museum	i-vii,9

N

National Library of Medicine	18
National Portrait Gallery	i-v
New York, State of	4,27,30,51,89,97,99
New York, New York City	103,105
North Star (periodical)	30,41,138

O

Ohio, State of	20,24,42,100,154

P

Pease, Joachim	49
Peculiar Institution	2,7,139
Pennsylvania, Chambersburg	41
Pennsylvania, Gettysburg, Battle of	105
Pennsylvania, Philadelphia	27,73,129,133
Physician	8,17-19

Pilot (wheelman)	33-34
Pinn, Robert	50
Planter	7
Plantation	1-3,7,27,59,89,93
Political (politician)	1-3,7,27,59,89,93,109,111,118,120
Poster	31
Pride	53,72,73,76,79,110,139,143
Pryor, Hubbard	58

Q

R

Race (racial)	17-18,22,28,32,42,91-92
Ratliff, Edward	50
Reconstruction (era)	9,33,42,91-92,101-102,106
Red Shirts	91
Revels, Hiram	iii, 100-102
Riot (Draft)	105
Ripley, Robert	8
Rivers, Prince	iii, 89-92

S

Seminary	100
Shaw, Mary	83
Shaw, Robert Gould Colonel	9,33,42,91-92,101-102,106
Siege	83,100,105
Slave (slavery)	1-14,22,29-32,59-60,90,94,102,104
Slave pen (auction house)	7,114,116

Slave trade	8,20,112,115,131-133
Smalls, Robert	33-35,161
Smith, Andrew	51
Smithsonian Institution	25
Social	2,28,122,126-127,143,151
South Carolina, State of	2-8, 13-17,33-47,51-74,81,133
South Carolina, 1st Colored Infantry Volunteers	35,40,82-83,89
South Carolina, Aiken	90-91
South Carolina, Beaufort	33,78,82-84,89
South Carolina, Charleston	14,33-35,41,74,85-86,89-90,127
South Carolina, Folly Beach	85
South Carolina, Hamburg	90-91
South Carolina, James Island	85,90
South Carolina, Morris Island	77,86
South Carolina, Port Royal	95
Speeches	6,23,29,107
Stanton, Edwin (Secretary of War)	35
Statesman	28,33,111
Stowe, Harriet Beecher	5,11-12,59
Supreme Court	3,134
Surgeon	15-19,22,40,42,59

T

Taylor, Susan Baker King	81,87-99
Thirteenth Amendment	32,102
Tillman, Juliana	5
Todd, Lockwood	23
Torture	59,117
Trowbridge, C.T. Colonel	86

Truth, Sojourner	iii, 1-28,154-157
Tubman, Harriett (Araminta Ross)	81,87-99

U

"Uncle Tom"	12,150
Uncle Tom's Cabin	11-12
Underground Railroad	26-27,29,42,94-95
Unknown (unidentified)	7, 25,58-60,71
U.S. Army	6,15-17,31-35,45,73-76,89-97,100
U.S. Colored Troops	6,19,31-33,50-51,59-60,80-87,89-97
29th Infantry Regiment, U.S. Colored Troops	80
33rd U.S. Colored Troops Infantry Regiment	35,82,85
4th U.S. Colored Troops Infantry Regiment	48,50-51,56
44th U.S. Colored Troops Infantry Regiment	58
73rd U.S. Colored Troops Infantry Regiment	60
107th U.S. Colored Troops Infantry Regiment	36,43
U.S. Navy	9,33-34,47-49,52-53,72-73,89,104
U.S. War Department	73
U.S.S. Hartford	47
U.S.S. Hunchback	53
U.S.S. Marblehead	47
U.S.S. Miami	52
U.S.S. Monitor	54
U.S.S. Wyandank	47

V

Veteran	87-88,97,99
Violence	3,12,21-22,30,81,91,141,143

Virginia	12,16-17,20,30,41,47,49-51,55,73,105,113
Virginia, Appomattox Court House	105-106
Virginia, Cumberland Landing	20
Virginia, Dinwiddie Court House	20
Virginia, Harper's Ferry (modern West Virginia)	12,30
Virginia, Richmond	49,105-106,127
Vogelsang, Peter	57
Vote	13,32,101,102,106

W

Washington D.C.	6,15,17,19-25,33-35,104-105
Watch Meeting	40
Wedgwood, Josiah	ix
"Whipped Peter"	59-60
White House	6,19-23,162
Wilberforce University	20,23-24,120
Wilson, James H.	57
Women	4,6,12,15,22,25,28,30,33,40,60,68,81,89,97

X

Y

Z

www.ingramcontent.com/pod-product-compliance
Lightning Source LLC
Chambersburg PA
CBHW081355290426
44110CB00018B/2385